STIQUITO™ for Beginners

STIQUITO™ for Beginners

An Introduction to Robotics

James M. Conrad
Jonathan W. Mills

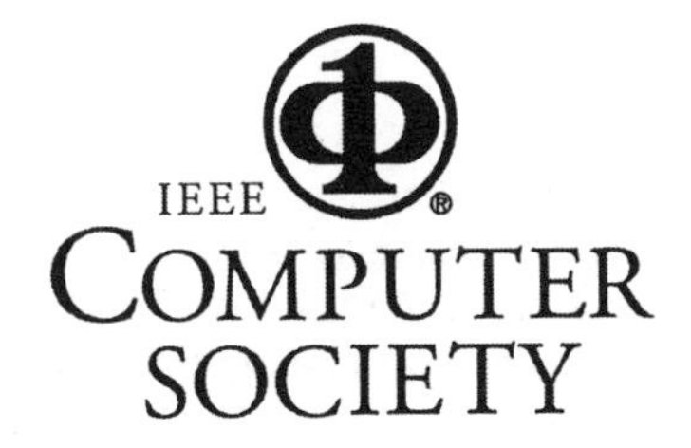

Los Alamitos, California

Washington • Brussels • Tokyo

Library of Congress Cataloging-in-Publication Data

Conrad, James M.
Stiquito for beginners : an introduction to robotics / James M. Conrad, Jonathan W. Mills.
p. cm.
Includes bibliographical references and index.
ISBN 0-8186-7514-4
1. Robotics. I. Mills, Jonathan W. (Jonathan Wayne). II. Title.

TJ211 .C636 1999
629.8'92 — dc21

99-052734
CIP

IEEE Computer Society Press Order Number BP07514
Library of Congress Number 99-052734
ISBN 0-8186-7514-4

Additional copies may be ordered from:

IEEE Computer Society Press
Customer Service Center
10662 Los Vaqueros Circle
P.O. Box 3014
Los Alamitos, CA 90720-1314
Tel: +1-714-821-8380
Fax: +1-714-821-4641
cs.books@computer.org

IEEE Service Center
445 Hoes Lane
P.O. Box 1331
Piscataway, NJ 08855-1331
Tel: +1-732-981-0060
Fax: +1-732-981-9667
mis.custserv@computer.org

IEEE Computer Society
Watanabe Building
1-4-2 Minami-Aoyama
Minato-ku, Tokyo 107-0062
JAPAN
Tel: +81-3-3408-3118
Fax: +81-3-3408-3553
tokyo.ofc@computer.org

Executive Director and Chief Executive Officer: T. Michael Elliott
Publisher: Angela Burgess
Manager of Production, CS Press: Deborah Plummer
Advertising/Promotions: Tom Fink
Production Editor: Denise Hurst
Printed in the United States of America

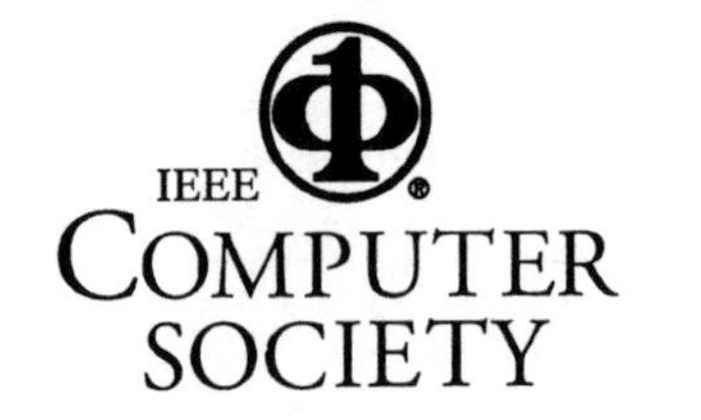

Contents

Preface

In 1992, Jonathan Mills announced the availability of Stiquito, a "small, inexpensive, hexapod robot." For $10 you could order from Indiana University a kit to build this small robot. Jonathan did not envision the number of requests he would receive, which by 1996 had reached more than 3,000. The volume of orders strained his personal ability to fulfill them; he started searching for alternate suppliers for the robot but had little success. In 1996, Jonathan was still receiving orders for Stiquito, although he had announced that Indiana University was no longer offering the kits.

In the summer of 1993 Jonathan began work on a research-oriented Stiquito book. In August 1994, I joined Jonathan in the effort to prepare and organize the chapters and identify other additional material that would be valuable in the book. By the fall of 1995, our publisher suggested that the original book be split into two books. The first book, *Stiquito: Advanced Experiments with a Simple and Inexpensive Robot,* was released in November 1997.

This second book is more educationally oriented than the first. This book has chapters more suited for high school and undergraduate students. It has the supplies needed to build one Stiquito robot. It also has an online teacher's manual, which includes additional experiments and lists the science benchmarks and national standards associated with each chapter and experiment. A solutions manual is also available from the publisher.

Stiquito for Beginners: An Introduction to Robotics can be used as the only book for an entire course, but it is best suited as one of several sources of material for a course. An early version of this book was used in an Introduction to Engineering Design and Skills course at North Carolina State University. It has been extensively reviewed by high school educators and university science education faculty.

This book is organized in three major parts: introductory material (Chapters 1–4), assembly instructions (Chapters 5–8), and summary and references (Chapter 9 and appendices). The specific topics of each chapter include:

- Chapter 1: An Introduction to Robotics and Stiquito—This chapter presents a brief overview of robotics and describes the Stiquito robot. This chapter also describes the skills needed to build a Stiquito robot.
- Chapter 2: Engineering Skills and the Design Process—What is an engineer? What skills does an engineer need? How does an engineer go about designing something? This chapter explores these questions.

- Chapter 3: Electricity Basics—This chapter discusses some of the basic properties of electricity and provides some experiments to show how electricity and electronic components work.
- Chapter 4: Nitinol Basics—This chapter discusses some of the basic properties of the Stiquito's muscle, nitinol wire. This chapter also provides some experiments to show how electricity and nitinol work.
- Chapter 5: Stiquito: A Small, Simple, Inexpensive Hexapod Robot—This chapter gives step-by-step instructions on how to assemble the robot kit included in the book.
- Chapter 6: A Manual Controller for the Stiquito Robot—You have finished building the robot kit. Now how do you make it walk? This chapter gives step-by-step instructions on how to build the simplest tethered controller for Stiquito.
- Chapter 7: A PC-Based Controller for the Stiquito Robot—This interface allows you to use an IBM PC or compatible computer to control the actuators on the Stiquito robot and experiment with various gaits.
- Chapter 8: A Simple Circuit to Make Stiquito Walk on Its Own—This chapter contains detailed instructions on how to build an electrical controller to allow Stiquito to walk autonomously.
- Chapter 9: The Future of Stiquito and Walking Robots—Now that you have built Stiquito, what can you do with it? What can it be used for in the future? This chapter explores these questions.
- A biography, a list of suppliers, and an index are also included.

How you use this book depends entirely on the course and time allocated for course material. Our suggestions are below. Refer to the Instructors Manual on the IEEE CS Press Stiquito web pages (http://www.computer.org/books/stiquito) for additional materials and suggestions:

- For an Introduction to Engineering course, follow this book sequentially. You will want to augment this book with other materials.
- For an electronics course, cover Chapters 3 (if needed), 5, 7, and 8 sequentially.
- For a robotics course (or to cover robotics within a course), use Chapters 1, and 3–9 sequentially.

As usual, we always recommend reading Chapter 5 before attempting to build the Stiquito Robot. If you read Chapter 5 completely before you build Stiquito, you will save time by avoiding common mistakes that everybody makes when trying to assemble the robot too hastily. You will also avoid wasting materials in your kit because of errors that the instructions can help you prevent.

Jonathan and I would like to thank the many people who assisted in the development of this book. Jon Butler, a volunteer of the IEEE Computer Society, provided continuous encouragement to Jonathan to create this book. Matt Loeb, Tom Fink, and Denise Hurst of IEEE Computer Society Press were instrumental in bringing this book to the marketplace. Ken Gracey of Parallax and Jon Pedersen of East Carolina University read drafts of the book and offered recommendations. Jamie Asbury of Hayssen and Bas Evers of Ericsson helped design, lay out, and test prototypes of the circuits and circuit boards described in Chapters 7 and 8. Murali Raju, Serge Caron,

Un Tung, Wesley Hisel, Larry Laxdal, Alexis Desbiens, and Tom Cooper tested the circuit in Chapter 8

Jonathan wishes to thank Indiana University and its Computer Science Department for the facilities they have provided for this project, notable in a liberal arts university that does not have an engineering school. He also wants to recognize his colleagues in the Computer Science Department, especially Steven Johnson and his students for their support and effort that turned Stiquito from what might have been just a toy into an educational and research tool that is in active use today.

I would like to recognize the University of Arkansas, especially my colleagues in the Computer Systems Engineering Department, and Collis Geren, Neil Schmitt, and Susan Vanneman for their support of Stiquito and engineering education. Special thanks go to Fritz Wilson and his team at Motorola University Support for their donation of Motorola semiconductors used in Stiquito education and research.

I also want to recognize my parents, the Conrads, my in-laws, the Warrens, and my children, Jay, Mary Beth, and Caroline, for their constant support during this project. I would especially like to thank my spouse, Stephanie Conrad, for her patience and support, even when deadlines dictated that I spend more time with Stiquito than with her.

James M. Conrad

Chapter 1

An Introduction to Robotics and Stiquito

James M. Conrad

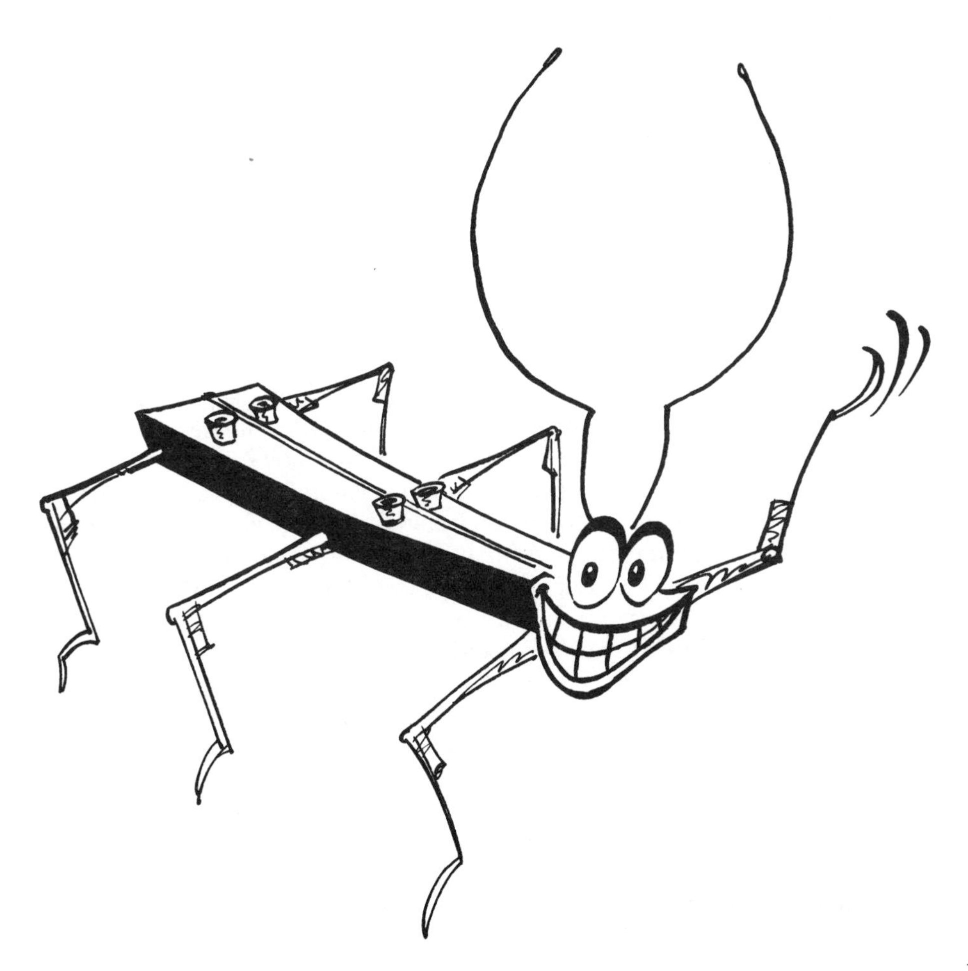

INTRODUCTION

Welcome to the wonderful world of robotics! This book will give you a unique opportunity to learn about this field in a way that has not been offered before. This book may also be the first affordable educational book to describe a robot **and** include the robot with the book! This book will provide you with the skills and equipment to build a small robot and with instructions on how to build electronic controls for your robot.

The star of this book is Stiquito, a small, inexpensive hexapod (six-legged) robot. Stiquito has been used since 1992 by universities, high schools, and hobbyists. It is unique not only because it is so inexpensive but because its applications are limitless.

This chapter will present an overview of robotics, the origin of Stiquito, and suggestions for how to proceed with reading the book and building the kit.

FIRST, SOME WORDS OF CAUTION

This warning will be given frequently, but it is one that all potential builders must heed. Building the robot in this kit requires certain skills to produce a working robot. These hobby building skills include:

- Tying thin metal wires into knots
- Cutting and sanding small lengths (4 mm) of aluminum tubing
- Threading the wire through the tubing
- Crimping the aluminum tubing with pliers
- Stripping insulation from wire
- Patiently following instructions that require 3 to 6 hours to complete

ROBOTICS

The field of robotics means different things to different people. Many conjure up images of R2-D2 or C-3PO-like devices from the *Star Wars* movies. Still others think of the character Data from the TV show *Star Trek: The Next Generation.* Few think of vehicles or even manufacturing devices, and yet robots are predominantly used in these areas. Our definition of a robotic device shall be: any electro-mechanical device that is given a set of instructions from humans and repeatedly carries out those instructions until instructed to stop. Based on this definition, building and programming a toy car to follow a strip of black tape on the floor is an example of a robotic device, but building and driving a radio-controlled toy car is not.

The term *robot* was created by Carl Capek, a Czechoslovakian playwright. In his 1921 play *RUR (Russums Universal Robots),* humans create mechanical devices to serve as workers.[1] The robots turn on their creators, thus setting up years of human versus machine conflicts.

The term *robotics* was coined by science fiction author Isaac Asimov in his 1942 short story "Runaround."[2] Asimov can be considered to be the biggest fan of robotics: he wrote more than 400 books in his lifetime, many of them about or including robots. His most famous and most often cited writing is his "Three Laws of Robotics," which he first introduced in "Runaround." These laws describe three fundamental rules that robots must follow to operate without harming their human creators. The laws are:

1. A robot may not injure a human being, or, through inaction, allow a human being to come to harm.
2. A robot must obey the orders given it by human beings except where such orders would conflict with the First Law.
3. A robot must protect its own existence as long as such protection does not conflict with the First and Second Laws.

These laws provide an excellent framework for all current and future robotic devices. There are many different types of robots. The classical robots depicted in science fiction books, movies, and television shows are typically walking, talking humanoid devices. However, the most useful and prevalent robot in use in the United States today is the industrial arm robot used in manufacturing. These robotic devices precisely carry out repetitive and sometimes dangerous work. Unlike human workers, they do not need coffee breaks, health plans, or vacations, although they do require maintenance and the occasional sick day. You may have seen an example of these robotic arms in auto maker commercials where an automobile body is welded and painted. Figure 1.1 shows an example of a small robotic arm manufactured by Seiko Instruments USA, Inc.

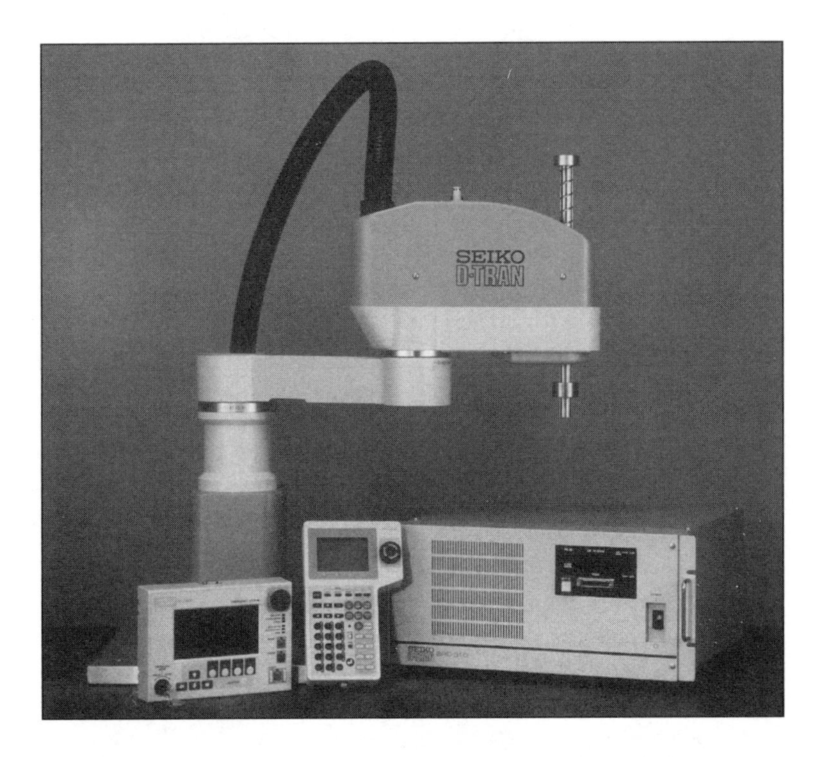

Figure 1.1. Robotic arm device. Used by permission of Seiko Instruments USA, Inc.

Another type of robot used in industry is the autonomous wheeled vehicle. These robots are used for surveillance or to deliver goods, mail, or other supplies. These robots follow a signal embedded in the floor, rely on preprogrammed moves, or guide themselves using cameras and programmed floor plans. An example of an autonomous wheeled robot, shown in Figure 1.2, is the SR 3 Cyberguard by Cybermotion.[3] This device will travel through a warehouse or industrial building looking for signs of fire or intrusion.

Figure 1.2. The SR 3 Cyberguard Autonomous Robot. Used by permission of Cybermotion, Inc.

Although interest in walking robots is increasing, their use in industry is very limited. Walking robots have advantages over wheeled robots when traversing rocky or steep terrain. One robot recently walked into the crater of a volcano and gathered data in an area too hazardous for humans to venture. Researchers at the University of Illinois have built a large walking robot, Protobot (Figure 1.3), based on the physiology of a cockroach.[4] Although humor columnist Dave Barry likens this 2-foot creature to a "FrankenRoach,"[5] the robot's designers envision such devices scurrying in hazardous environments and even adapting to the loss of a limb.

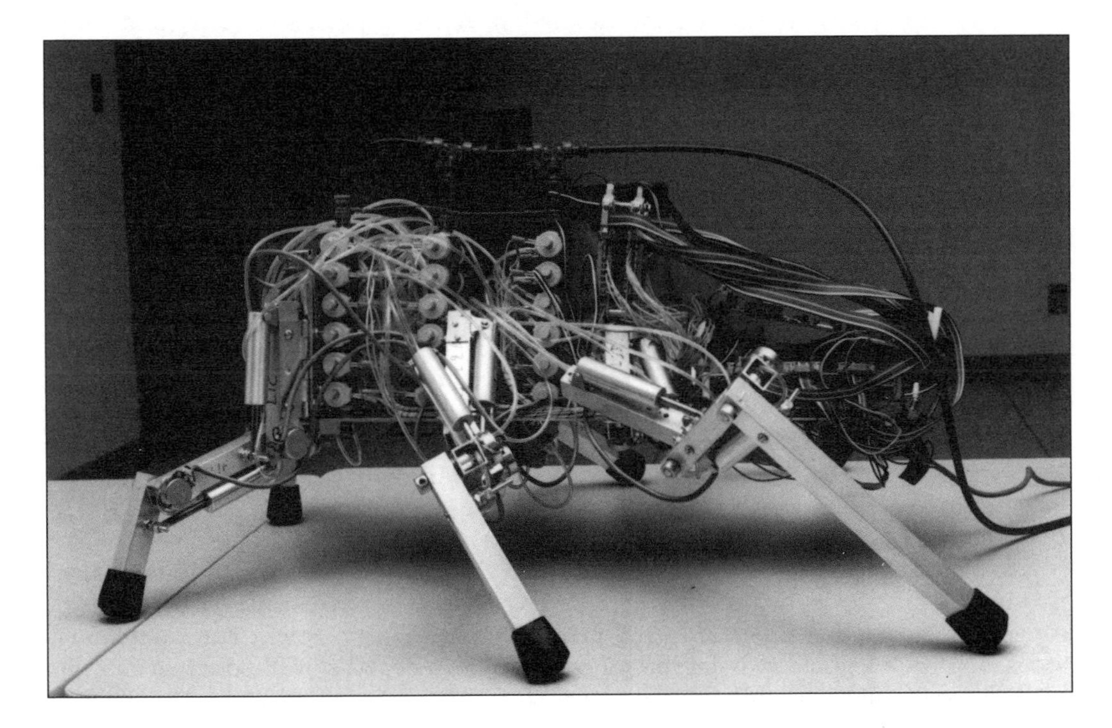

Figure 1.3. Protobot, a cockroach-inspired walking robot. Used by permission of IEEE CS Press.

Most walking robots do not take on a true biological means of propulsion, defined as the use of contracting and relaxing muscle fiber bundles. The means of propulsion for most walking robots is either pneumatic air or motors. Protobot approaches a biological construction because it walks by means of pneumatic cylinders that emulate antagonistic muscle pairs.

True muscle-like propulsion did not exist until recently. A new material, nitinol, is used to emulate the operation of a muscle. Nitinol has the properties of contracting when heated and returning to its original size when cooled. An opposable force is needed to stretch the nitinol back to its original size. This new material has spawned a plethora of new small walking robots that originally could not be built with motors. Although several of these robots were designed in the early 1990s, one of them has gained international prominence because of its low cost. This robot is called Stiquito.

STIQUITO

In the early 1990s, Dr. Jonathan Mills was looking for a robotic platform to test his research on analog logic. Most platforms were prohibitively expensive, especially for a young assistant professor with limited research money. Since necessity is the mother of invention, Dr. Mills set out to design his own inexpensive robot. He chose four basic materials from which to base his designs:

- For propulsion, he selected nitinol (specifically, Flexinol™ from Dynalloy, Inc.). This material would provide a muscle-like reaction for his circuitry and would closely mimic biological actions. More detail on nitinol is provided in Chapter 4. Other sources contain detailed specifications on Flexinol™.[6]
- For a counterforce to the nitinol, he selected music wire from K & S Engineering. The wire could serve a force to stretch the nitinol back to its original length and provide support for the robot.
- For the body of the robot, he selected ⅛-inch square plastic rod from Plastruct, Inc. The plastic is easy to cut, drill, and glue. It also has relatively good heat-resistive properties.
- For leg support, body support, and attachment of nitinol to plastic he chose aluminum tubing from K & S Engineering.

Dr. Mills experimented with various designs, from a tiny four-legged robot 2 inches long to a floppy six-legged, 4-inch long robot. Through this experimentation he found that the best movement of the robots was realized when the nitinol was parallel to the ground, and the leg part touching the ground was perpendicular to the ground.

The immediate predecessor to Stiquito was Sticky, a large hexapod robot. Sticky is 9 inches long by 5 inches wide by 3 inches high. It contains nitinol wires inside aluminum tubes. The tubes are used primarily for support. Sticky can take 1.5 cm steps, and each leg had two degrees of freedom. Two degrees of freedom means that nitinol wire is used to pull the legs back (first degree) and raise the legs (second degree).

Sticky was not cost effective, so Dr. Mills used the concepts of earlier robots with the hexapod design of Sticky to create Stiquito (which means "little Sticky"). Stiquito was originally designed for only one degree of freedom but has a very low cost. Two years later, Dr. Mills designed a larger version of Stiquito, called Stiquito II, which had two degrees of freedom.[7] A picture of Stiquito II is shown in Figure 1.4.

At about the same time that Dr. Mills was experimenting with these legged robots, Roger Gilbertson of MondoTronics and Mark Tilden of Los Alamos Labs were also experimenting with nitinol. Gilbertson and Tilden's robots are described in our first Stiquito book.

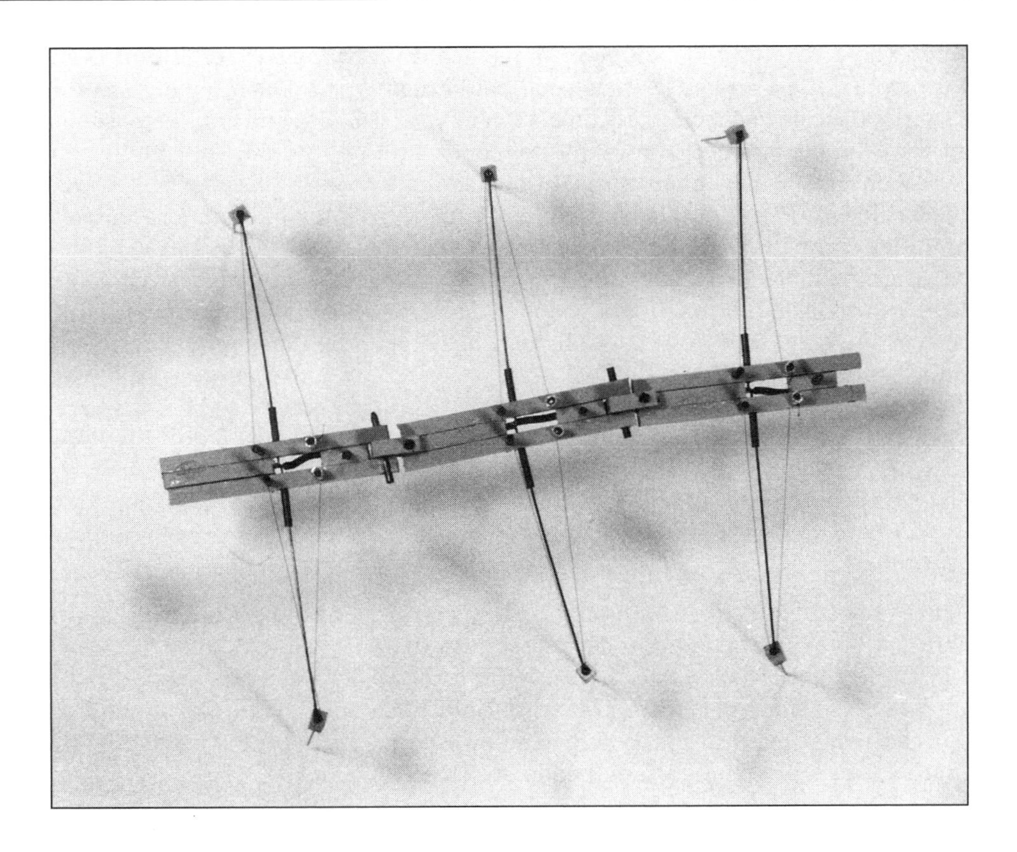

Figure 1.4. The Stiquito II robot.

THE STIQUITO KIT

The kit included with this book has enough materials to make one Stiquito robot, although there are extra components in case you make a few errors while building the robot. The most important thing to remember when building this kit is that Stiquito is a hobby kit; it requires hobby-building skills, like cutting, sanding, and working with very small parts. For example, in one of the steps, you will need to tie a knot in the nitinol wire. Nitinol is very much like thread, and it is very difficult to tie a knot in it. However, if you take time and be patient (and after some practice), you will soon be able to tie knots like a professional.

The kit included with this book is a simplification of the original Stiquito described in Dr. Mills' technical report[8] and offered as a kit from Indiana University. In your kit the plastic Stiquito body has been premolded, so now you no longer have to cut, glue, and drill the plastic rod to make the body. Because of this simplification, more than 10 pages were removed from the original instructions. This new body also allows builders to make more errors, and requires less precision when building the robot; therefore, your robot should be more robust than earlier models.

The intent of this kit is to allow builders to create a platform from which they can start experimenting. The instructions provided in Chapter 5 show how you can create a Stiquito that walks in a tripod gait; that is, it allows three legs to move at one time. What you should do is to examine your goals for building the Stiquito and plans for controlling how Stiquito walks. If your plans include allowing each of Stiquito's six legs to be controlled independently, then you should modify the assembly of your robot to attach control wires to each leg. If the design of your robot includes putting something on top, for example a circuit that will allow it to walk on its own, you should consider how you want it to walk. If you want it to walk simply (as in Chapter 6), a tripod gait may be sufficient. If you plan to put some complex circuitry like a microcontroller on top, you may want the flexibility to control all six legs.

The Stiquito robot body was also designed so you can assemble it using screws instead of aluminum crimps. If you wish to use screws instead of crimps, use the sets of holes on the body that are offset slightly. See Appendix C for an example of using screws in building Stiquito.

The Stiquito body is also designed so that all 12 large holes can be used for allowing the legs to have two degrees of freedom (just like Sticky and Stiquito II). You cannot control this robot with the manual controller, but you can control it with the PC parallel port card described in Chapter 7.

Chapter 5 has very detailed assembly instructions, but here are some additional handy hints:

- This is not Lego®. Stiquito is not a snap-together, easy-to-build kit. As a hobby kit, it takes some model building skills. Be patient! Allow 6 hours to build your first robot. Dr. Mills swears he can build a robot in 1 hour, but it takes me about 3 (while watching sports on TV). This could be a wonderful parent-child project (in fact, my elementary school-aged son wants to "build bugs with daddy"). Make sure to block out enough time to complete the kit.
- Make sure you do not introduce any shorts across the control and ground wires on the robot, tether, or manual controller. Feel free to use electrical tape to insulate areas that might cause a short.
- Make sure all electrical connections are clean and free of corrosion. Sand metal parts before tying, crimping, or attaching.
- You may need to add some weight to Stiquito when using the manual controller. You can tape pennies to the bottom of the body or tape a AA battery on top. Make sure the weight does not short the control and ground wires.

In your building activities, we cannot stress the importance of following common safety practices:

- Wear goggles when working with the kit. Many parts of the kit act as sharp springs!
- Use care when using a hobby knife. Always cut away from you.
- Use care when using a soldering iron. Watch out for burns!

This kit is intended for adults and for children over the age of 14.

EDUCATIONAL USES OF STIQUITO

One of the main functions of this book is to provide an environment for educators to introduce robotics and robotic control to students. This book can be used at the high school or university level to introduce students to the concepts of analog electronics, digital electronics, computer control, and robotics. Since this book comes complete with assembly instructions and a robot kit, it can easily serve as a required textbook for a class with only a minimal amount of electronics required to investigate the other areas.

The Stiquito robot also provides opportunities to learn about other engineering disciplines. For example, previous exercises using Stiquito have included creating Stiquito kits similar to what was found in the back of the book from raw products. Given long lengths of music wire, long lengths of aluminum tubing, spools of wire, and bags of other parts, fill a small plastic bag with the parts necessary to build Stiquito later. This requires students to contemplate Industrial Engineering concepts, including assembly lines and materials handling. A description of this exercise can be found on the IEEE Computer Society Press's Stiquito web page at: http://computer.org/books/stiquito.

As with any project conducted in a school setting, you will need additional supplies for those cases when students break their robot kit. Contact the IEEE Computer Society Press to purchase additional kits, or contact some of the suppliers listed in the back of the book for repair materials.

WHAT NEXT?

Now that you have read the Preface and Chapter 1, you should have a good idea of the contents of this book. You need to consider your goals with respect to Stiquito. If you are a student and this book is assigned by your instructor, this process may have been already determined (but you can always do more!). Use the chapters of the book

Figure 1.5. A favorite Stiquito builder cartoon.

that help you attain your goals. Remember, you can always buy more kits from the publisher (IEEE CS Press) if your goal is to build more than one robot.

EXERCISES

1. Consider the three Laws of Robotics.
   - Can you think of a situation where the three Laws of Robotics would fail in their intent?
   - Do the three Laws of Robotics take into account the competing goals of good and evil?
2. Read the Isaac Asimov short story "Runaround." This story can be found in many compilations of his short stories like *The Complete Robot* or *I, Robot.* Check your school or local library for these books. Why did the robot Speedy fail?
3. List all the robots you have seen. Identify if these are manufacturing robots or autonomous robots. Use the definition given in this chapter to categorize these machines.
4. Muscle-like contractions of wire propel the Stiquito robot. Can you think of any other machines that have muscle-like motion?
5. Search your library or the World Wide Web for information about insectoid robots. List the types of insects that engineers are trying to replicate.
6. Most successful robots perform a small range of activities very well. Identify a small, real-world problem that could be solved with a new robot. Design a robot to solve the problem. Use your imagination. Some solutions invented by students include:
   - A woodpecker-like robot to drive nails in inaccessible surfaces
   - A snake-like robot to traverse drain pipes and clear clogs
7. Today's successful mobile robots typically use wheels for movement. For example, Cyberguard and mail/food delivery robots use wheels. Of course, in nature there are no wheels.
   a. Identify the advantages and disadvantages of wheeled autonomous robots.
   b. Identify the advantages and disadvantages of legged autonomous robots.
   c. Which type of robot would work best on flat terrain? Which would work best in rough, sloped terrain? Which would work best indoors? Which would work best in a football stadium? Justify your answers.
   d. Based on your answers to parts a, b, and c, which approach holds the most promise for future applications?

BIBLIOGRAPHY

1. Capek, C. 1972. *Russums universal robots* (translated and reprinted.)
2. Asimov, 1982. *The complete robot.* Garden City, N.Y.: Doubleday.
3. SR 3 Cyberguard. Cybermotion, Inc., home page. Available: http://www.cybermotion.com

4. Price, D. 1995. Climbing the walls. *IEEE Expert* (April): 67–70.
5. Barry, D. 1995. Dr. FrankenRoach. *The Washington Post Magazine,* p. 32.
6. Dynalloy, Inc. *Technical characteristics of Flexinol™ actuator wires.* See also Appendix D of *Stiquito: Advanced experiments with a simple and inexpensive robot.*
7. Conrad, J. M., and J. W. Mills. 1997. *Stiquito: Advanced experiments with a simple and inexpensive robot.* Los Alamitos, Calif.: IEEE Computer Society Press.
8. Mills, J. W. 1992. Stiquito: A small, simple, inexpensive hexapod robot. Technical Report 363a, Computer Science Department. Bloomington, Ind: Indiana University.

Other Sources

Gibilisco, S. 1994. *The McGraw-Hill illustrated encyclopedia of robotics & artificial intelligence.* New York: Mc Graw-Hill.

Rosheim, M. E. 1994. *Robot evolution: The development of anthrobotics.* New York: Wiley and Sons.

Chapter 2

Engineering Skills and the Design Process

James M. Conrad and Allan R. Baker

INTRODUCTION TO ENGINEERING

The impact of engineering on your life is tremendous. Engineering achievements surround you every day. Every building, bridge, road, automobile, and computer were designed by engineers. Engineers designed the manufacturing plants that made the automobiles and computers. Simply, nearly everything that is manufactured was designed by engineers.

What is an engineer, and what is engineering? The American Society for Engineering Education (ASEE) defines engineering as:

> *The art of applying scientific and mathematical principles, experience, judgment, and common sense to make things that benefit people. Engineering is the process of producing a technical product or system to meet a specific need.*

An engineer is a professional who uses engineering and design processes to make things that benefit people.

The field of engineering includes several major disciplines. The discipline and the products or processes they design and implement are:

- **Chemical**—chemical processes for the production of petrochemicals, fertilizer, rubber, plastic, processed foods, and fuels.
- **Civil/Environmental**—roads, bridges, buildings, dams, sewage treatment plants, water treatment plants, and other structures considered infrastructure.
- **Computer**—computer programs and computer systems (see Figure 2.1).
- **Electrical/Electronic**—electrical generation and distribution networks, electronic devices, integrated circuits, radio devices, and communications devices.
- **Industrial/Manufacturing**—production lines, quality control, ergonomics, and logistics.
- **Mechanical**—machine enclosures and machines, including: automobiles, airplanes, boilers, and power plants.
- **Systems Engineering**—system modeling, system analysis, system requirements, and system testing, or making sure the mechanical, chemical, electrical, and computer components of a system work together to do the required job.

Some of the smaller engineering disciplines include agricultural, biological, biomedical, materials, nuclear, and petroleum engineering.

Figure 2.1. A computer engineer testing software on a new cellular phone.

Stiquito relies on knowledge from several of these disciplines. A robot may be an inherently mechanical device, but its movement, or locomotion, is based on electricity. The nitinol used as the main muscle of Stiquito was engineered with materials and chemical principles. Computer engineering skills are used to program it to walk. The design of a specific walking gait uses biological insight.

The ASEE reports that 1.2 million engineers were working in the United States in 1998. According to the U.S. Labor Department, job growth in engineering over the next few years should range from 110% in computer engineering to 5% in petroleum engineering.[1] This trend also extends worldwide. Refer to the ASEE web site for more detail on current salaries.[2]

What do you need to do to prepare for an engineering career? Most entry-level jobs require at least a bachelor's degree in engineering. To prepare for college, you should take as many courses as possible in math and science. Most successful college students earned good grades during 4 years of math and 4 years of science in high school. By no means do you need to be a genius, but you should have a solid mastery of the subject you studied. After all, engineering applies the art of mathematical and scientific principles.

Persistence is an important trait for an engineer, because many successes are achieved through hard work and experience. Although an engineer works more efficiently as he/she gains experience, a passion for the subject matter and enthusiasm will help the engineer learn and gain that experience.

Note that scientists also use mathematical and scientific principles in their work. What distinguishes an engineer from a scientist is the end product. Scientists work primarily to expand a body of knowledge in their field, while engineers use that knowledge to create things. For example, a pharmaceutical chemist may investigate how a particular new chemical compound reacts with blood. A chemical engineer will use the chemical knowledge to create processes to manufacture the new chemical compound. In addition, while scientists use the scientific method in their work, engineers use the engineering design process in their work.

ENGINEERING DESIGN

Engineering design is the creative process of identifying needs and then devising a solution to fill those needs. This solution may be a product, technique, structure, project, method, or other result, depending on the problem. The general procedure for completing a good engineering design can be called the *engineering method of creative problem solving.*

Problem solving is the process of determining the best possible action to take in a given situation. The types of problems that engineers must solve vary between and among the various branches of engineering. Because of this diversity, no universal list of procedures will fit every problem. Not every engineer uses the same steps in the design process. The following list includes most of the steps that engineers use.[3]

1. Identify the problem.
2. Gather the needed information.
3. Search for creative solutions.
4. Overcome obstacles to creative thinking.
5. Move from ideas to preliminary designs (including modeling and prototyping).
6. Evaluate and select a preferred solution.
7. Prepare reports, plans, and specifications (project planning).
8. Implement the design (project implementation).

Although this list of steps is usually followed, note that not all engineering efforts use all steps. Some projects may even use a subset of the steps in a different order. Perhaps a project started out as a hobby or a university research project. The designer may have found that the product could be marketed as it is designed, thereby skipping steps like searching for creative solutions or evaluating and selecting the preferred solution.

Before an engineering effort begins, sales, marketing, and engineering managers examine the economics of the project. The product developed by engineers must be marketable, and the cost of developing the project should be much less than the total revenue derived from product sales. Typically, this group creates a business case using current trends or market research to establish a need for the product. Engineering management estimates the cost of development and production, and marketing and sales create sales forecasts and pricing. If a product will not be profitable, the project is usually not started.

Of course, not all projects require a business case. Some projects are so important, they must be developed no matter the cost. An example might be ensuring that a database system recognizes dates with the year 2000 and beyond, an endeavor which occupied many software engineers in the late 1990s.

Identifying the Problem

Identifying the problem is the first and most important step in finding a solution. If the problem is not defined properly, an engineer will likely waste time or arrive at an incorrect solution. It is important that stated needs be real needs. A great design may be worthless if it duplicates existing designs or does not benefit many people.

When defining the problem, the engineer must be careful to understand the needs completely and not limit the range of possible solutions. Take for example the recent 1997 NASA mission to Mars. In the government era of cutting all types of spending, the most challenging problem was no longer just getting to Mars. The new challenge was *cost.* How could NASA go to Mars cheaply?

In the past, NASA engineers had the luxury of huge budgets and many years of development time. "Better, better, better" was the attitude that drove everything they did. If the old attitude had been used when identifying the problems of putting a lander on the Martian surface for the Pathfinder mission, they might have ended up with a craft much like the Viking landers that made the trip in the 1970s. Viking used parachutes and rockets to guide the spacecraft gently to the surface. It was crucial that the delicate scientific instruments not be damaged in the plunge through the thin Mars atmosphere. With the old attitude and a proven design in mind, their requirement might have been very specific.

> **Definition 1:**
> *The lander's retro rockets must have a controlled thrust capability of reducing the speed of a 500 kg craft from 1000 km/s to 5 km/s within a distance of 500 m with an overall force less than 6 g.*

Contrast this with the vague requirement below:

> **Definition 2:**
> *The lander must be able to survive Martian reentry speeds while not exposing the internal instruments to deceleration forces that would damage them.*

The first definition gives the design team specific requirements toward which they may work. The second definition might suffice for some situations, but for this situation, it gives too much leeway to the design team. Using this definition, the finished product may not accurately meet the needs of the customer.

On the other hand, notice how the first definition limits the range of possible design solutions. Its requirements limit the team to using rockets to slow the craft's fall, a manner that may be difficult to meet.

For the NASA of the '90s the new requirements driving production are "faster, better, cheaper."[4] First, the lander and rover had to be cheap to build. Second the engineers needed to be quick to design and build. Finally, they had to be better than previous spacecraft. Unlike the "better, better, better" motto of the past, it was painfully obvious to NASA that the agency's future was bleak if cost was not given top priority. Traditional methods may have met the physical performance criteria, but not the fiscal performance criteria. For their design, NASA engineers used a combination of parachutes, small retrorockets, and inflatable airbags that allowed the lander to release its kinetic energy by bouncing across the surface of Mars.

Gathering Needed Information

After defining the problem, an engineer begins to gather all the information and data necessary to solve the problem. It could be physical measurements, maps, results of laboratory experiments, patents, results of opinion surveys, or other types of information.

Engineers should always try to build on what has been done before. Information on related problems that have been solved or unsolved may help engineers find the best solution. In the NASA example, engineers used consumer technology that was already available to meet costs and performance requirements. To design things from scratch would have been extremely expensive.

Searching for Creative Solutions

Several techniques are useful to help a group or individual to produce original creative ideas. The development of these new ideas may come from *creativity,* a subconscious effort, or *innovation,* a conscious effort.

Groups or individuals can use several techniques to come up with creative solutions, including brainstorming, checklists, attribute listing, and the forced relationship technique.[5] Each of these will be briefly discussed. The purpose of most of these methods, which can be almost like a game, is to break the set patterns of thought that every individual develops.

Brainstorming—Brainstorming is popular technique for group problem solving. Typically, a brainstorming session consists of 6 to 12 people who spontaneously suggest ideas for solving a specific problem. In these sessions, participants are encouraged to record all ideas, including those that appear completely impractical. Ideas should not be judged or evaluated during brainstorming sessions. What is most important is to generate as many ideas as possible and encourage people to build upon the ideas of others. Ideas should be evaluated after the session is completed. Individuals can also use the same techniques to brainstorm without a group.

BRAINSTORMING RULES

1. Encourage all ideas.
2. Record as many ideas as possible.
3. Combine and improve ideas.
4. Delay judgment and evaluation of ideas until end.

Checklists—One of the simplest methods for generating new ideas is to make a checklist. The checklist encourages the user to examine various points, areas, and design possibilities.[6] For example, a checklist like this might be developed in the process to improve existing NASA exploration craft:

Ways the device could reuse existing commercial technology

Ways the device could be made lighter

Ways the device could be simplified

Ways the device could be made cheaper

Ways the device could be made with less design time

Attribute Listing—With attribute listing, all the major characteristics or attributes of a product, object, or idea are isolated and listed. Then, list ideas on how to change each of the attributes. Again, as in brainstorming, all ideas are listed no matter how impractical. After all the ideas are listed, evaluate each idea for possible improvements that can be made to the design.[7] For example, how can we improve on a spacecraft design?

Attribute	**Ideas**	
1. Size	Could be smaller Could be split into multiple parts	Could be lighter Could be inflatable
2. Atmospheric decent	Could use glide wings Could bounce	Could use parachutes Could use rockets
3. Lander motivation	Could use wheels Could use legs	Could be stationary Could use wings
4. Power and propulsion	Could use solar energy Could use solar wind	Could use nuclear energy Could use fly-by gravity assists

Forced Relationship—The forced relationship technique takes a fixed element, such as the product or some idea related to the product, and forces it to take on the attributes of another unrelated element. This forms the basis of a free flowing list of associations from which (hopefully) new ideas will emerge. As before, judge the value of the ideas after the process is complete.[8]

For example, suppose we wish to use this technique on our Martian lander. This will be the forced object. Suppose we randomly choose an automobile wheel as the other element. Some of the ideas that may occur based upon the automobile wheel are:

A round Martian lander

A rubber Martian lander

A Martian lander that rolls

A Martian lander that has spokes

A Martian lander that has air in its tires

A Martian lander that has brakes

A Martian lander that will not break

Overcoming Obstacles to Creative Thinking

Sometimes, the more experience we have with an existing solution, the more difficult it is to be creative and to generate new ideas for a different and possibly better solution. Some specific actions and attitudes can be employed to overcome obstacles to creative thinking.[9,10]

1. Don't overconstrain the problem. In other words, do not make the problem appear too difficult to solve.
2. Seek out different viewpoints.
3. Diversify your team. Ten people with the same idea is still only one idea.
4. Avoid preconceived beliefs and stereotypical thinking.
5. Realize that many problems have non-engineering solutions. Consider approaches that other disciplines might use.
6. Embrace the weird, unusual, and untraditional. Most creative thought involves putting experiences and thoughts into new patterns and arrangements.
7. Simplify complex problems. Divide large problems into pieces that are more manageable, and concentrate on one part at a time.
8. Relax and reflect after periods of intense concentration. Many creative thinkers have shouted out "Eureka!" as they realized the solution to their problem while taking a bath or showering in the morning.
9. Be open-minded about problem-solving strategies and solutions.

Turning Ideas into Designs

To turn their ideas into a design, engineers must sort through lots of possible solutions and determine which ones are promising and which ones are not. The promising ideas are then molded and worked into plans. Preliminary designs evolve through *analysis* and *synthesis.* Engineers analyze by breaking apart the whole and studying its individual components. They synthesize by putting together many pieces of information into a whole idea that will accomplish their goal.

An engineer can choose from many techniques to determine if an idea has promise. An idea may be physically the best possible solution, but may be too costly. As competition increases and great ideas and solutions are plentiful, cost is sometimes the overwhelming deciding factor in choosing an idea.

Obsolescence, or the state of gradually becoming antiquated, is also a big deciding factor for technology solutions. It seems technology has a half-life of about 5 years; that is, half of today's technology products will be obsolete in about 5 years. In 1993, an Intel 66MHz 486 computer was a good buy. In 1999 the 300MHz Pentium® IIs were soon to be replaced as top product by the next generation of microprocessors. The possibility of a design becoming obsolete could be a big issue if the product being designed today does not hit the market for 3 or 4 years.

A preliminary sketch or analysis may show that the idea is a bad one. Research may be needed to determine the advantages and disadvantages of each design choice. Tests may need to be run on components to see if they will work in extreme environments. Perhaps a large research project may be needed to examine the validity of a process and the consequences if it is used as a solution. The engineer must be critical when studying the possible solutions and constantly eliminate poor or inappropriate solutions.

To facilitate the design process, engineers often rely on models. A model simplifies a system or process so that it may be better studied, understood, and used in a design. Four common models are used in engineering: mathematical, simulation, physical, and prototype.

Mathematical models usually consist of one or more equations that describe a physical system. Many physical systems can be described by mathematical models. Such models can be based on scientific theories or laws that have stood the test of time, or they may be based on empirical data from experiments or observations. Mathematical models are usually employed for simple systems, although advances in computers, machine intelligence, desktop spreadsheets, and commercial simulation tools have expanded the usefulness of mathematical models. For example, an aeronautical engineer can design an airplane wing and use a fluid dynamics model to predict the airflow across the wing.

Computer simulation models allow engineers to examine complex systems. Such models may incorporate many mathematical models as part of the total simulation model. A computer program is developed to describe a system, and this model can be run repeatedly under many different simulated operating conditions.

Today, electrical engineers use computer simulation models to design digital (and in some cases even analog) circuitry. These models can fully simulate the function of circuits without ever programming a programmable logic device (PLD) or fabricating an application specific integrated circuit (ASIC) using commercial simulation software tools. An example of an electrical engineer using a circuit simulation product is shown in Figure 2.2.

Figure 2.2. An electrical engineer simulating a new cellular phone circuit design.

Developments in genetic algorithms, artificial neural networks, and other techniques now give researchers the ability to define the desired result of a model and let the computer find the best solution. Complex data from real-world systems can be input to a computer so that the computer can "learn" how to generate similar outputs for similar inputs that it may or may not have seen before. Fuzzy logic algorithms are another development that allow models to be constructed based on "English" descriptions of the models instead of mathematical ones.

Physical models have long been used by engineers to understand complex systems. They probably represent the oldest method of structural design. Physical models have the advantage in that they allow an engineer to study a device, structure, or system with little or no prior knowledge of its behavior. Full-scale models are sometimes built, but most often, they are scaled down anywhere from 1:4 to 1:48. Examples of studies made with physical models include:

1. Dispersion of pollutants throughout a lake
2. Behavior of waves within a harbor
3. Underwater performance of different submarine shapes
4. Performance of aircraft by using wind tunnels to simulate various flight conditions

Prototype models are used in addition to other modeling to prove that a design works, test the synthesis of a complex design, work out bugs, or make tougher design decisions when the models are not accurate enough. In prototyping the engineer attempts to build a fully functional device based on the initial designs, versus a physical model that is not to full size or is not fully functional.

A prototype for a complicated design need not include the entire design. Software can be tried out in a simulation of computer hardware or software. Electrical designs can be built by hand on breadboards or inexpensive wire wrapping tools. NASA built mechanical mock-ups of its shuttle and space station designs so that it could test how well the pieces fit together and turn up any ergonomic or manufacturing problems that would not show up on paper or in a computer simulation.

Prototyping pieces of a design so that a complex problem is broken into smaller, more manageable problems has many advantages.

Evaluating and Selecting a Preferred Solution

Engineers use several criteria to evaluate the value of a solution or design, depending on the nature of the problem. If the solution involves a product, great importance may be placed on safety, cost, reliability, and consumer acceptability.

Many designers use prototypes to test the operation of the design. The designer could then identify any weak areas of the design and attempt to improve upon them. No idea should be discarded solely based on one prototype or one test. Many great designs have been discarded prematurely and many working prototypes have failed to give acceptable products. Then again, many designs that were thought to be great were actually flawed. One only needs to look at the explosion of the space shuttle Challenger or the collapse of the Tacoma Narrows Bridge for an example.[11,12] In both of these disasters, serious design flaws caused the destruction of the spacecraft and structure.

Indirect evaluation can also be used to evaluate a design. For example, scale models can be used to test aircraft design at a fraction of the cost of building a prototype.

In this example, computer simulations and mathematical models may not be accurate enough to allow an engineer to understand all the complexities of component interference or turbulence, but they still may be used to approximate the design of the first scale model for wind tunnel testing.

Preparing Reports, Plans, and Specifications

After selecting a design, it must be shared with those who must approve it, support it, and translate it into reality. This communication may take the form of an engineering report, or a set of plans and specifications. Engineers use plans and specifications to describe to a manufacturing division or to a contractor the details about a design so that it can be produced. Engineering drawings, written and oral communications, and scheduling and planning a design project are essential in implementing a design smoothly and efficiently. Some materials engineers use to support their design plans include engineering drawings, written communication, oral communication, and design project schedules.

Engineering Drawings—Engineers often create detailed technical drawings that show what the design looks like, what parts are necessary, how to assemble it, and how to operate it once constructed. These *graphical specifications* are probably the most important type of documentation used for engineering design problems. They communicate visually to the technical team what verbal communications cannot adequately convey. To be effective, these drawings must be drawn clearly and according to standards and conventions accepted by the team.

Written Communication—Memorandums, often called memos, are a brief and effective way to keep everyone involved aware of the design's progress. Memos can be distributed to one person or to a list of people within the organization who have an interest in the subject.

A technical report is a much longer and complete record of the design process. It should include everything that was done to solve the problem. As with any communication, the technical writing should be clear, direct, and readable by the intended audi ence. Many types of reports are written by engineers, but in general, they all include the following information:[13]

- Cover page, stating title of project, company name, author, and date
- Abstract, giving a short overview and summary of the work
- Table of contents
- Body of the report, which elaborates the problem, presents background material, procedure used to solve the problem, results, and significance of work
- Conclusions and recommendations, which summarize the results and significance of work
- Appendices for supplementary material not needed in the main body of the report

Oral Communication—At different stages during the design process an engineer may be called upon to give an oral progress report to the design team, supervisor, management, or marketing staff. The objectives of an oral presentation are the same for a written report: the engineer wants to communicate information about the project. The methods used, however, are very different. The most important element for a successful oral presentation is preparation. Here are some additional pointers for a good oral presentation:

- Be very familiar with the subject of the presentation.
- Know how much time is allotted for the presentation.
- Practice the presentation to cover everything completely within the time limit.
- Know your audience. Match presentation level with the audience's understanding level.
- Speak clearly and eloquently.
- Have simple and uncluttered visual materials. Do not put too much information on one visual.
- Have a good summary and conclusion to highlight the important parts of the presentation.
- End the presentation by asking the audience if they have any questions.

Design Project Schedules—Since a design project is usually much more complicated than finishing physics homework, a complete solution will involve several steps or tasks. Some complex problems will require weeks or months to complete. The solution to the design problem needs to include a schedule or a plan of when certain tasks in the design should be completed. In a good schedule, each task is completed before its results are needed by another task. The schedule should also use all personnel all the time. Since designs frequently are changed and improved throughout the process, it is a good idea to schedule design reviews at the end of each project phase. An example of a project plan is shown in Figure 2.3.

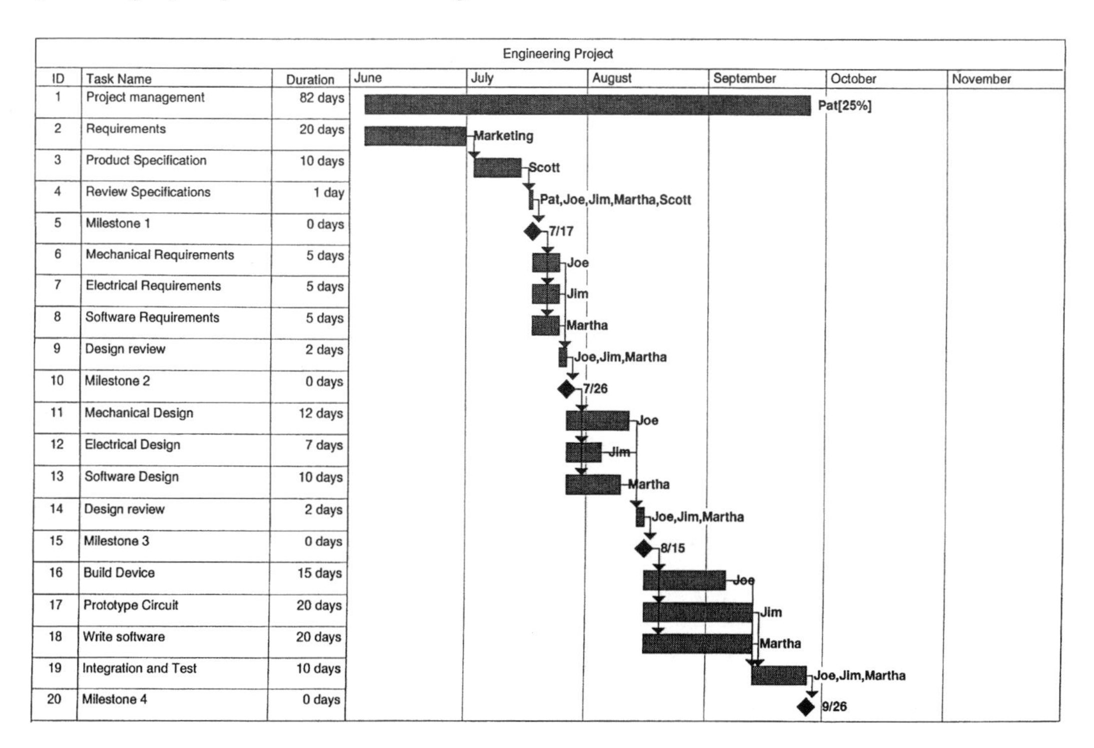

Figure 2.3. A sample project plan of an engineering development cycle.

Implementing the Design

The final stage of the design process is implementation, or the process of producing or constructing a physical device, product, or system. Engineers usually plan and oversee the production and construction of the engineered projects. The design engineer may not necessarily be the engineer involved in the final phase. For the design engineer, implementing the design is the most satisfying stage of all.

CASE STUDY: ENGINEERING DESIGN

Creating a Stiquito is one example of the engineering design process. Chapter 1 examines the history of Stiquito. A continuation of the history shows how the original design, or more precisely, the Stiquito body, was modified and included in the current kit.

Identifying the Problem

Consider the original Stiquito as the prototype for the current Stiquito kit. The original Stiquito had several shortcomings as a commercial product. Users told us that building the robot required several hard-to-get tools, drilling small precise holes was difficult, and many metal parts touched each other. We identified our problem definition as:

> *Design the Stiquito body as one piece that needs no subassembly or drilling and allows 1/16″ more room between metal parts.*

We discussed this problem with the publishers of the book, and all decided that to make the book a commercial success, the body should be one complete piece.

Gathering Needed Information

To examine how to solve the problem, we gathered information about the existing body.

From the original Stiquito specification:[14]

1. The original body required that two pieces of plastic be cut and glued together.
2. The glued pieces were drilled by hand; six holes ⅛-inch diameter and six holes 1/64-inch diameter were needed.
3. The drilling was required to be precise.

Information was gathered from our existing users:

1. The glue sometimes was not strong enough to hold the two parts together.
2. The objects inserted into the body had very little room for error, since the body was only ¼-inch wide. Electrical shorts were frequent.

Searching for Creative Solutions

One day Jonathan Mills and Jim Conrad sat down and brainstormed about how to solve the problem. Possible solutions for a new Stiquito body included:

- Single piece of molded plastic
- Single piece of molded plastic with the holes in it
- Single piece of plastic with holes, and with the legs and metal backbone molded in the plastic
- Single piece of molded plastic with the holes in it, plus extra holes for other things
- A printed circuit board
- Leave it as is

Discussions then continued by way of the telephone and e-mail.

Moving from Ideas to Preliminary Designs

Some of the ideas described in our process were implemented. For example, a new Stiquito design with the body being a printed circuit board was made and tested. We even asked a printed circuit board company to submit a quote for the body. Our tests showed this solution did not work.

Another attempt was to "simulate" a single piece of plastic by drilling and gluing three pieces of plastic together. This seemed to work.

To make a commercial product, we proposed that a single piece of plastic be provided so the user would not need to do any cutting, drilling, or gluing to the plastic body. A good engineering process examines the shortcomings of a design and then makes modifications to the design. After our observations were noted, we made the following design decisions:

1. The body would be a single piece of plastic, using some sort of plastic manufacturing process; it would have approximately the same dimensions as if the pieces were glued together.
2. The piece of plastic would come complete with predrilled holes or otherwise created with the same dimensions.
3. By making the body itself ⅛ inch wider, the builder of the Stiquito has more room for error. In other words, the design enables the user to not have such narrow tolerances.

We determined that the best way to create the plastic body was using a mold and making a plastic injection molding of the body. A plastic mold would also be used for the manual controller.

A prototype was created to examine how this would work and if it were possible. A mold maker at the University of Arkansas made a quick, rough mold and produced about 50 pieces to test. The mold was chosen more for speed than durability.

Evaluating and Selecting a Preferred Solution

During the design process, we also found that sometimes people had difficulty clipping the nitinol in the aluminum. Another design was created to allow screws rather than aluminum clips. The designer concluded that to wrap nitinol around the screws, you needed to take into account the direction of the threads. The new design of the Stiquito body has holes that would ensure that the nitinol would wrap in the direction of the threads.

Preparing Reports, Plans, and Specifications

Once the prototype went through several different versions of tests, the design was finalized. A drawing was made of the Stiquito body and the manual controller as shown in Figure 2.4.

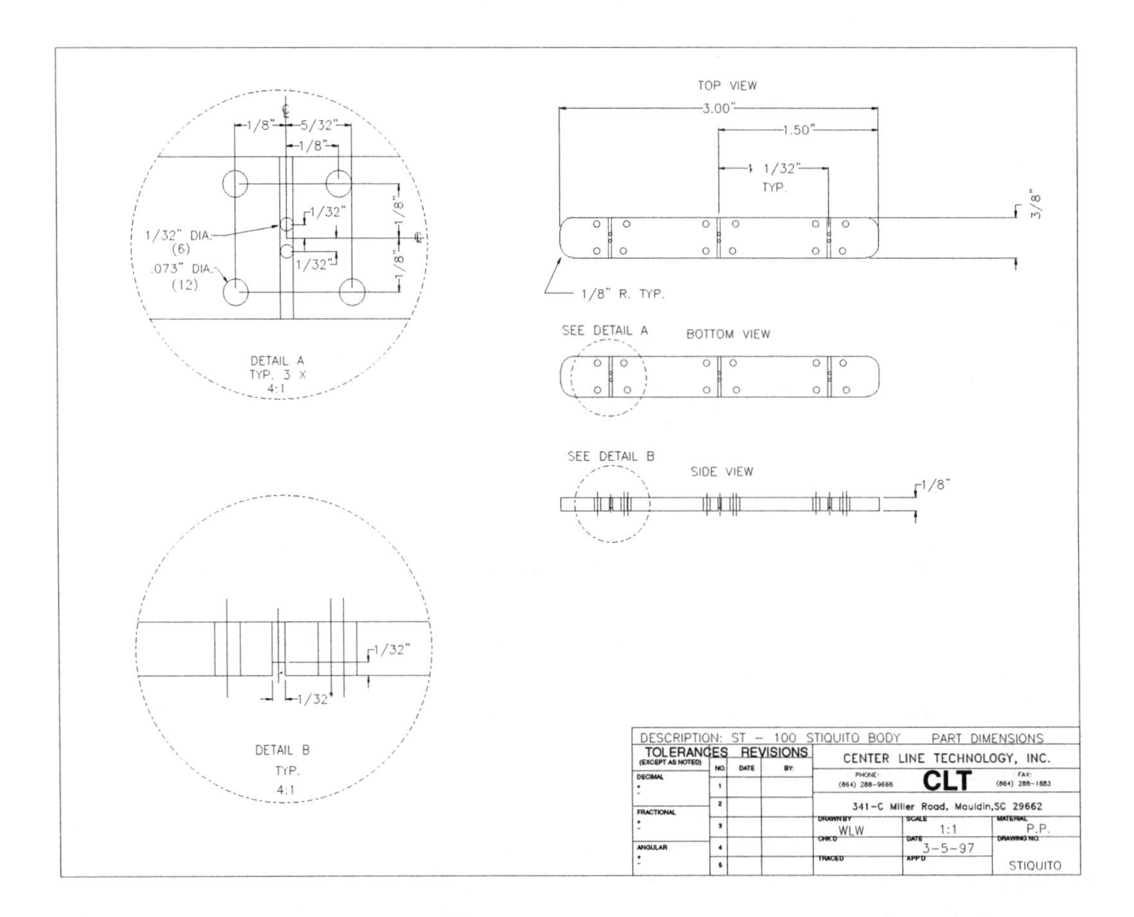

Figure 2.4. The Stiquito design document, drawn using mechanical design software.

Bids were solicited from mold makers and plastics companies, and vendors were selected. The mold maker required more information than the designers specified to make a mold. Very often when creating a product, the designer and manufacturer must continue to communicate beyond passing a drawing from one to the other. In our case, the vendor needed information not represented in the design to ensure the robot was manufactured correctly. The additional information included:

1. What were the tolerances of the widths and lengths of the Stiquito robot?
2. For the ease of manufacturing, could the corners be rounded rather than perfectly square?
3. Again, for ease of manufacturing, could the sides be tapered 2 degrees to allow the finished product to be removed from the mold more easily than if the sides were perfect 90 degrees with the top and bottom?
4. Could the pins used to make the holes in the mold be tapered to allow the device to be easily removed?
5. What was the desired final finishing of the device?
6. Was any lettering desired for the device? (We added lettering for copyright.)
7. What was the estimated production volume for this device? (This was important because a harder, more expensive metal could be used if the expected volumes were going to be high.)
8. Who was the intended manufacturer of the device? Certain manufacturers had plastic injection parts, molds, and machines that could make less expensive molds by taking advantage of standard tools.
9. What was the intended material? (The intended material for the body affected the heat transfer, removal, and cooling required for typical plastic injection molding.)

Implementation

Once the mold was created, it was tested at the manufacturer for the plastic injection. Based on the first runs through the manufacturing process, it was determined that the specified hole sizes were too large and should be smaller to allow the wrapped wire and aluminum to fit more snugly in the hole. A picture of the mold is shown in Figure 2.5.

Figure 2.5. The Stiquito plastics mold.

Case Study Conclusion

Based on the engineering design process, the problem was identified, a design was made, and prototypes were created. A manufacturing engineer identified additional questions that affected the original design, and the original design was modified. The designer remained involved until the product was manufactured. In the engineering process, often an engineer will maintain contact or remain responsible for future upgrades, changes, or modifications to the original design. In our example, the design was straightforward and relatively simple. The changes that were made were minor and did not drastically affect the original intent of the device. This is not always the case in the engineering design process. The more complex an object or system and the more engineers involved in the project, the more communication increases and the more chance there is for a major design change that affects how the product functions.

A successful engineer continues to learn throughout his or her career. In the case of the new Stiquito body, the designers had to learn about routing and pricing a printed circuit. We also were introduced to the world of plastics manufacturing. Our success was based on following good design principles and relying on experts for advice.

REAL-WORLD EXAMPLES

To further demonstrate the design process, we present two case studies of products that James Conrad developed at BPM Technologies and Ericsson, Inc.

The first case study is on a project at BPM Technology (which is no longer in business). BPM designed and manufactured a device that would convert computer-aided design drawings into a three-dimensional piece of plastic (see Figure 2.6). This "printer" used an ordinary PC as its user interface and also had a PC interface card that sent instructions to the printer hardware.

The designer of the PC interface card did not follow the typical design steps, and the card needed to be modified. James Conrad served as the project manager and lead technical developer. He created design documents and had other designers review the documents. More work and features were added to the project, and the schedule started to extend. People who were supposed to work on the project were assigned other work, so the delivery date extended further in the future. Eventually, the hardware and software were implemented, but the company closed its doors in the middle of testing!

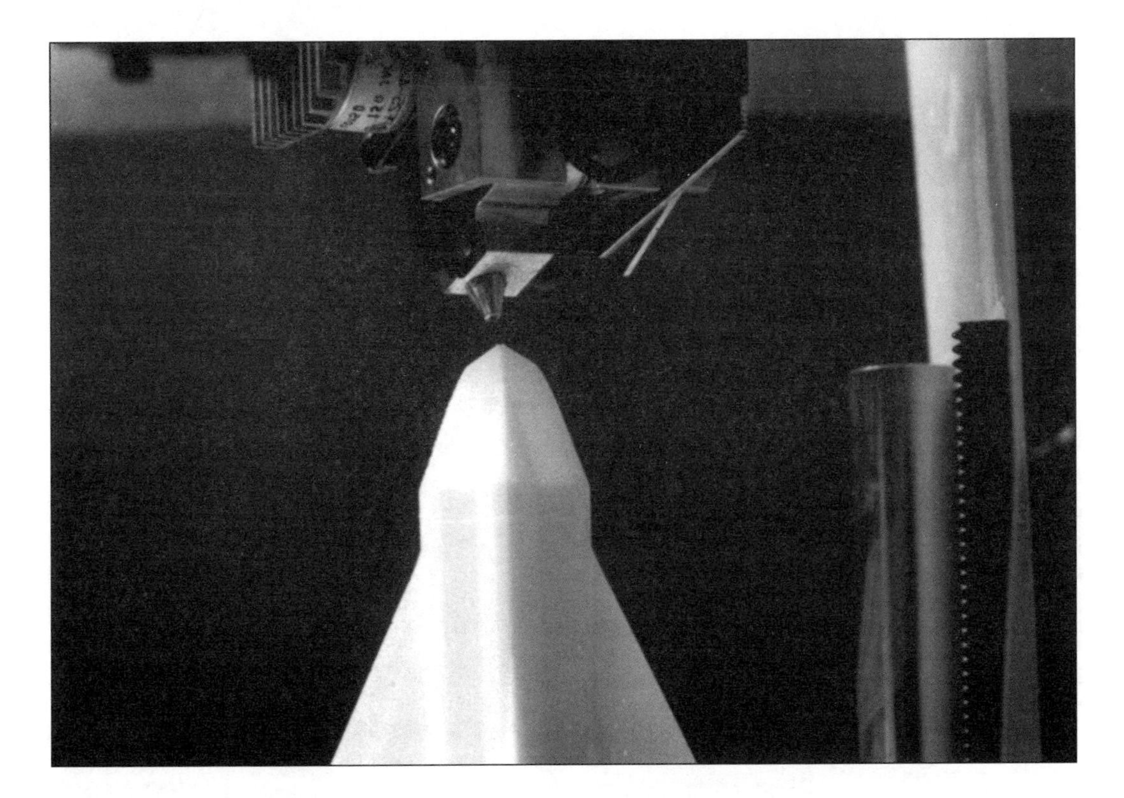

Figure 2.6. The BPM three-dimensional printer.

The second case study is on a code reuse project completed at Ericsson and described in more detail in a paper presented at the 1999 Engineering of Computer-Based Systems Conference.[15] The product developed was a variation of three digital cordless telephone products previously developed by Ericsson's Holland and Research Triangle Park laboratories. The new product (see Figure 2.7) used designs and code from all three existing products. The new product was entirely a software reuse effort. All hardware needed was already available from the previously developed system or from other Ericsson products.

In compliance with ISO9001 standards, the Ericsson development team followed a company methodology to complete this project using the Ericsson Software Development Process. This methodology used the following steps:

- Business case development and management support
- Requirements analysis
- Global design

Figure 2.7. The Ericsson DCT1900 Freeset Indoor Cellular Phone System.

- Detailed design
- Implementation
- Component and integration test
- Function and system test

Sticking to the formal process, relying on the strengths of the prior products, and tapping the experience of their teams provided the developers a clear sense of the scope of the work and the current state of progress. This created a solid focus on the goal, and the team worked hard to achieve it. The best measure of the success of the project was the result: the product was delivered on time and fully conformed to the specification.

EXERCISES

1. What is the difference between a scientist and an engineer?
2. What type of engineer would design:
   a. A coat hanger?
   b. A light bulb?
   c. A canal?
   d. A video game?
   e. A cellular telephone?
3. What are the eight steps engineers use for creative problem solving?
4. Brainstorming is a technique for what creative problem-solving step? What are the rules of brainstorming?
5. What process can an engineer perform to gain more knowledge and information about the intended design for a product? Why is this important?
6. In the case study of the new Stiquito body design, what was done in the preliminary design step?
7. Identify a problem that needs a solution. Do not choose a large problem, just something small, like a better mousetrap. Using the eight creative problem-solving steps, create your better mousetrap.
8. Interview an engineer about his or her educational background, work, and design process. Find out how the engineer first became interested in engineering.

BIBLIOGRAPHY

1. Basta, N. 1996. *Opportunities in engineering careers.* Lincolnwood, Ill.: VGM Career Horizons.
2. American Society for Engineering Education (ASEE). 1999. http://www.asee.org/precollege
3. Wright, P. H. 1994. *Introduction to engineering,* 2nd ed. New York: Wiley & Sons, pp. 91–117.

4. Pritchett, P., and B. Muirhead. 1998. *The Mars Pathfinder approach to "faster-better-cheaper."* Dallas: Pritchett & Associates, Inc.
5. Wright, 1994.
6. Wright, 1994.
7. Wright, 1994.
8. Wright, 1994.
9. Wright, 1994.
10. VanGundy, A. B. 1982. *Training your creative mind.* Englewood Cliffs, N.J.: Prentice Hall.
11. ASEE, 1999.
12. Petroski, H. 1992. *To engineer is human: The role of failure in successful design.* New York: Vintage Books.
13. Howell, S. K. 1995. *Engineer's toolkit, a first course in engineering: Engineering design and problem solving.* New York: Benjamin/Cummings Publishing, pp. 65–70.
14. Mills, J. W. 1992. Stiquito: A small, simple, inexpensive hexapod robot. Technical Report 363a, Computer Science Department. Bloomington, Ind.: Indiana University.
15. Conrad, J. M., M. Baldwin, S. Curran, and L. Martin. 1999. Using a new software product development process for a code reuse project. *Proceedings of the 1999 Engineering of Computer-Based Systems Conference,* Nashville, Tenn., March, pp. 34–40.

Other Sources

Beakley, G. C., and H. W. Leach. 1982. *Engineering: An introduction to a creative profession,* 4th ed. New York: Macmillan.

Macaulay, D. 1998. *The way things work.* Boston: Houghton Mifflin.

Simon, H. A. 1975. *A student's introduction to engineering design.* New York: Pergamon Press, pp. 100–101.

Chapter 3

Electricity Basics

James M. Conrad

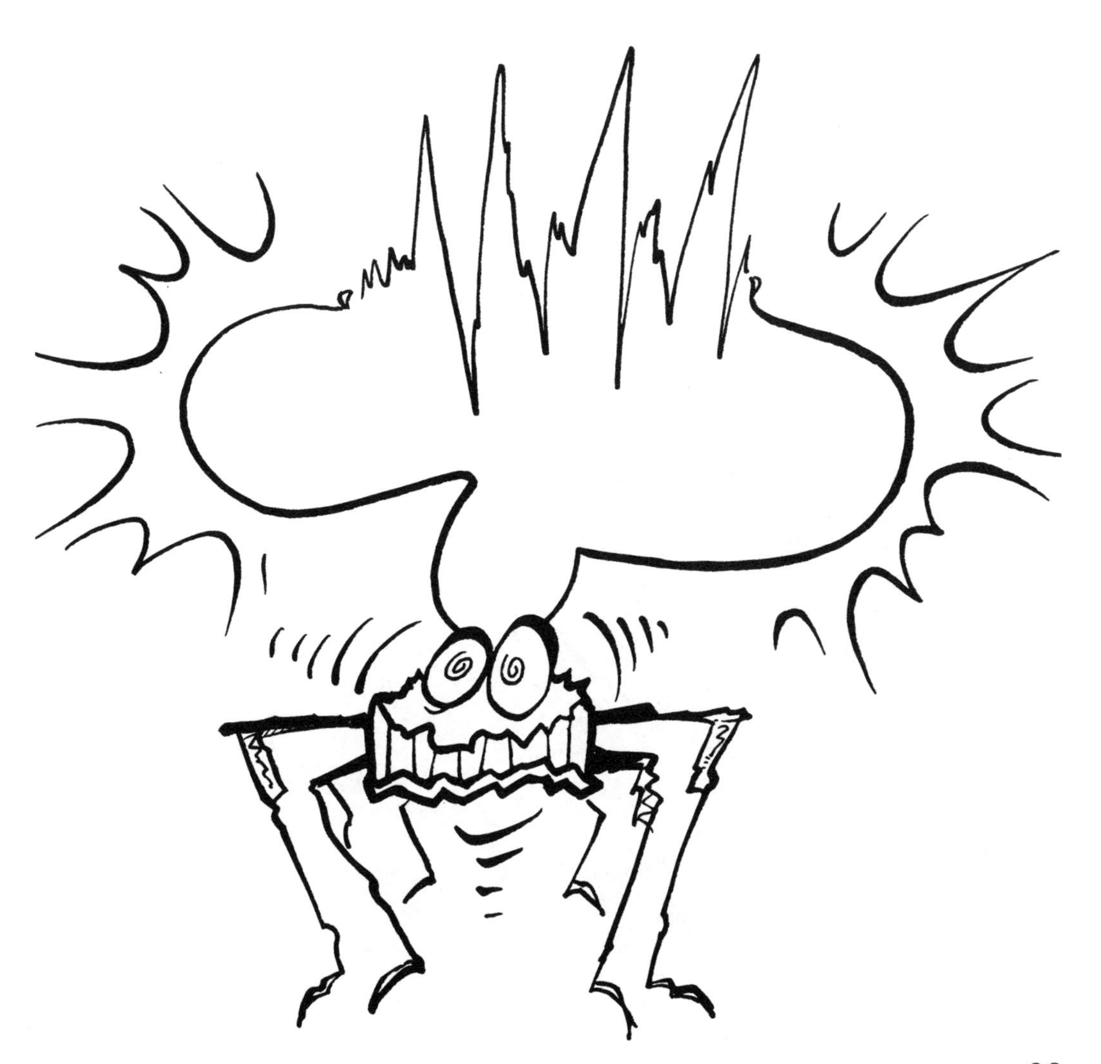

INTRODUCTION

This chapter includes a brief introduction to the basic concepts of electricity. Also discussed are the electronics you will need to build Stiquito and operate it using a manual controller, a digital personal computer controller, or an analog controller. The digital controller and the analog controller must be purchased separately, although the instructions to build them are included in this book.

Electricity was probably one of the most important resources of the 20th century. One need only imagine the problems associated with having no electricity to see how essential it is and how much we rely on it. For example, should you lose electrical power to your home due to some natural disaster, you may find that you have no heat, air conditioning, lights, or chilled food in your refrigerator. Depending on the appliances in your home, you may have no stove to cook food, and perhaps no water because electricity is needed to pump water into your home.

ELECTRICAL CIRCUITS

To understand electricity and electrical circuits, we first must define three terms: voltage, current, and resistance.

First, think of an electrical circuit such as a battery and a light bulb. This circuit is shown in Figure 3.1. Once the circuit is connected, electrons flow from the negative (minus) side of the battery, through the light bulb to the positive (plus) side of the battery. As the electrons flow, they provide **current** to the light bulb, causing it to light. Current is the movement or flow of an electric charge. This example shows that a current will flow between two oppositely charged bodies, namely the positive and negative sides of a battery. Current is measured in *amperes,* also called *amps.*

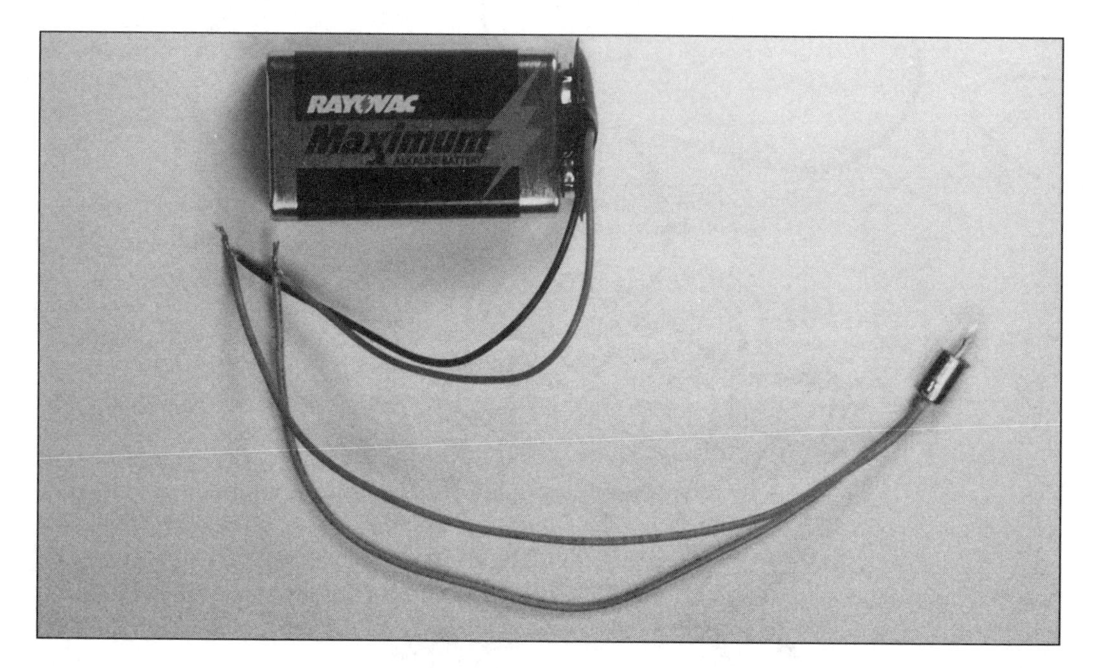

Figure 3.1. A simple battery and light bulb circuit.

Simply put, electricity is the flow of electrons from a high potential to a low potential. A battery is an example: current flows from the positive side of the battery (high potential) to the negative side (low potential). The difference in these potentials of the battery is called **voltage.** Conductors transport these electrons along their solid structure. The conductors we work with are most often copper and aluminum wires. Voltage is measured in *volts.*

Although electricity flows in copper wires mostly unimpeded, there still is some resistance to its flow. **Resistance** is defined as the force needed to impede the flow of electrons. Most often a byproduct of this resistance, or slowing of the electrons, is heat; sometimes it is light, and sometimes it is an electromagnetic force that creates a magnetic attraction. As a simple example, consider the light bulb and battery. The battery serves as the source of the electrical current, and the light bulb serves as the resistor. As the electrical current flows through the light bulb, it generates heat. In the case of the light bulb, the result of this heat is a very bright light. Resistance is measured in *ohms* and is also represented by the Greek letter omega (Ω).

Resistance causes a voltage drop across a resistor as current flows through it. In any circuit, there is a direct relationship between current, voltage, and resistance. The German physicist Ohm identified this relationship. Ohm's law is expressed as $V = I * R$, where V is the voltage in volts, I is the current in amps, and R is the resistance in ohms.

You can always find the value of one of these terms if the other two are known. By simple algebra, Ohm's law can also be expressed as $I = V/R$ and $R = V/I$.

When using values of volts, amps, and ohms, it is often necessary to use the prefixes commonly used for metric measurements. These prefixes are shown in Table 3.1.

Table 3.1. Prefix Abbreviations

Prefix	Value
M (Mega-)	× 1,000,000
K (Kilo-)	× 1,000
m (milli-)	× 0.001
μ or u (micro-)	× 0.000001
n (nano-)	× 0.000000001
p (pico-)	× 0.000000000001

BEFORE YOU BEGIN EXPERIMENTS . . .

Remember that the electricity used in your school, home, or work is dangerous; if you are careless, it can be fatal. We have greatly eliminated this risk by using only small voltages (3 to 9 volts). Still, you should follow some basic safety precautions:

- Always wear safety glasses when performing this work.
- Although the electrical components used in this chapter are not particularly static sensitive, avoid handling them by the pins.
- In winter and dry weather, touch a large metal object (table, door frame) to discharge any static electricity before handling electrical components.

The people who design electrical circuitry typically draw their circuits before they build them. These designers use different symbols to represent each part, such as the resistance. Figure 3.2 shows several of the symbols used.

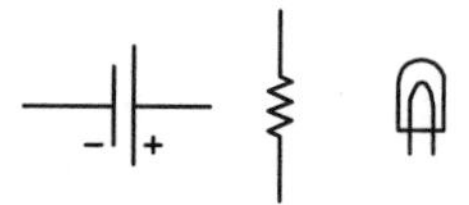

Figure 3.2. Circuit drawing symbols: Battery, resistor, light bulb.

Materials Needed

To complete the experiments in this chapter you will need the following materials, which are illustrated in Figure 3.3. These items are contained in the Stiquito Educational Kit (not the Stiquito Robot Kit) and are readily available from your local electronics store:

- One 9-volt battery connector
- One light bulb with wires (180 ohms at 9 volts)
- One ¼ watt, 1K ohm resistor
- Two ¼ watt, 100K ohm resistors
- One integrated LED (LED and resistor in one package)
- One 2.2 microfarad, 50 volt capacitor
- One 2N2222 transistor
- One momentary switch
- One length of wire wrap wire

Refer to the section "Reading Resistor Values" at the end of this chapter to distinguish correctly the 1K ohm and 100K ohm resistors.

You will also need the following materials and tools that are not contained in the Stiquito Educational Kit:

- 9-volt battery
- Electrical tape

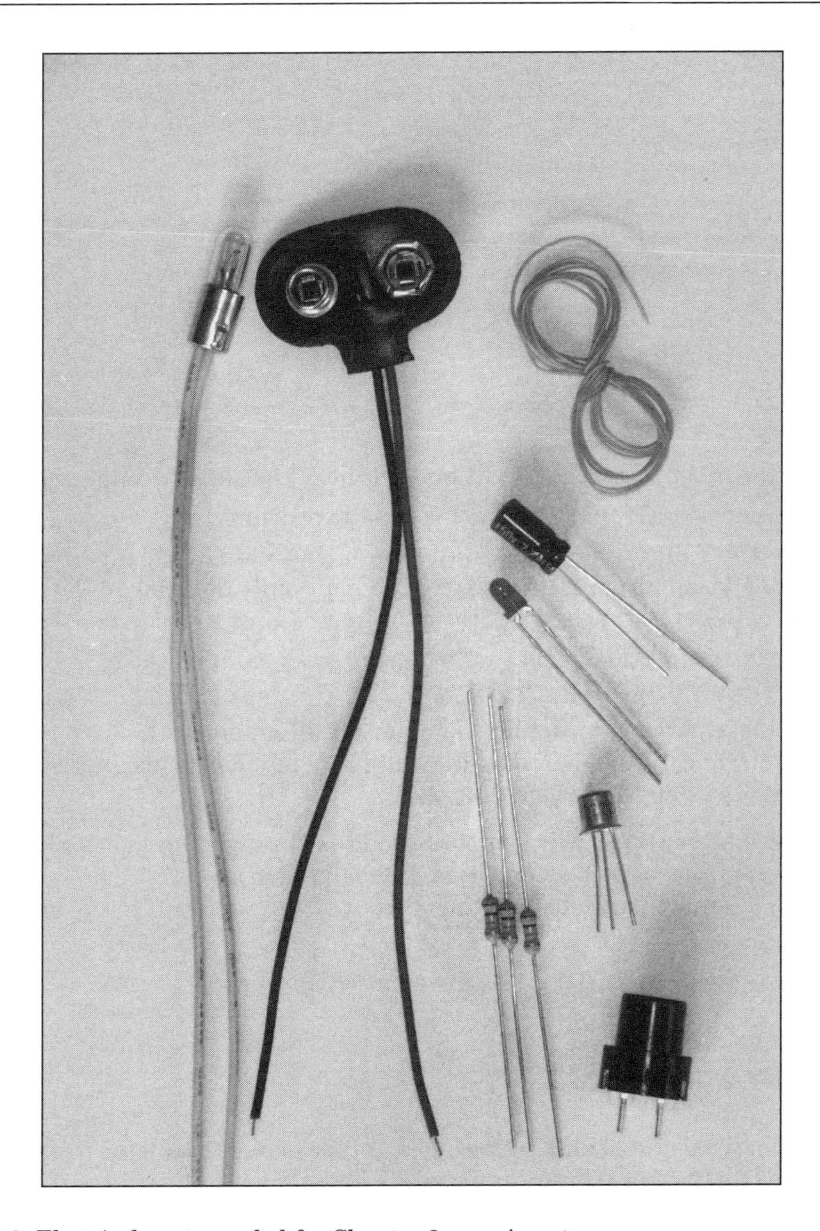

Figure 3.3. Electrical parts needed for Chapter 3 experiments.

EXPERIMENT 1: A SIMPLE CIRCUIT

This experiment examines the circuit shown in Figure 3.1. It demonstrates how electricity flows from a simple battery to a resistive light bulb.

Experiment 1 requires the following materials:

- 9-volt battery
- 9-volt battery connector
- Light bulb with wires

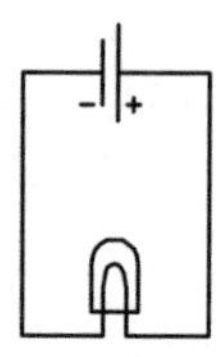

Figure 3.4. A simple battery and light bulb circuit.

1. Attach one end of the incandescent bulb wire to one wire of the 9-volt battery connector.
2. Attach the other wire of the light bulb to the other battery connector wire.
3. Attach your 9-volt battery to the 9-volt battery connector.
4. Observe the light and its brightness. You have created a closed circuit in which current is flowing from the battery to the light bulb and back to the battery. The battery actually has a chemical reaction that creates this current flow.
5. Remove the connection between the battery connector and the light bulb, remembering which side was attached to which.
6. Attach the other side of the light bulb to the first side of the battery connector, and then connect the other side. Do you notice any difference in the intensity of the light? Does anything look different?

It does not matter which way the current flows through the light bulb; the same heat will be generated and the same light will be produced.

Using Ohm's law, we can calculate the current of our circuit. If $V = 9$ volts, $R = 180$ ohms, then:

$I = V/R$ = 9 volts/180 ohms = 0.050 amps = 50 milliamps

EXPERIMENT 2: RESISTORS

Very often in the world of electronics another type of device called a resistor is used. Resistors limit the flow of current in a circuit. Experiment 2 illustrates this concept.

Experiment 2 requires the following materials:

- 9-volt battery
- 9-volt battery connector
- One ¼ watt, 1 K ohm resistor

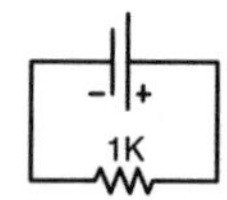

Figure 3.5. A simple resistor and light bulb circuit.

1. Take the resistor and attach the ends to the battery connector as shown in Figure 3.5.
2. Attach the 9-volt battery to the battery connector. Do you see any difference or change in the resistor?
3. Carefully touch the resistor. Does it feel any warmer than when you first hooked it up?
4. Using Ohm's law, calculate the current flowing through the resistor.

This electronic resistor, does not produce any light, but it does produce heat. Although it does not generate enough heat to heat your classroom or your house, this type of device serves a purpose in the world of electronics. It controls the amount of electrical current allowed to flow from the battery into a circuit and back into the battery.

EXPERIMENT 3: SERIES RESISTORS

This experiment shows what happens to the current of a circuit when two resistive devices are attached in series. Series means the current flows through the circuit one resistor at a time. Experiment 3 illustrates this concept.

Experiment 3 requires the following materials:

- 9-volt battery
- 9-volt battery connector
- Light bulb with wires
- One ¼ watt, 1 K ohm resistor

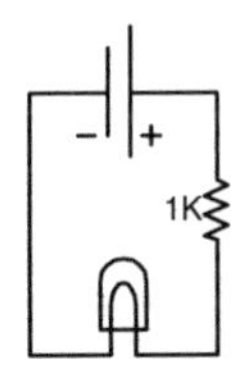

Figure 3.6. A light bulb and a resistor in series.

1. Take the resistor and attach the ends to the battery connector as shown in Figure 3.5.
2. Attach the 9-volt battery to the battery connector. Do you see any difference or change in the resistor?
3. Carefully touch the resistor. Does it feel any warmer than when you first hooked it up?
4. Now add the light bulb to the circuit, as shown in Figure 3.6. Does the light bulb look brighter or dimmer than in Experiment 1? You may want to connect the circuit in Experiment 1 again to compare the brightness.

This experiment shows what happens when you have two resistive loads in series. In this case, Ohm's law is represented by $V = I * (R1 + R2)$, where *R1* is the resistance of the resistor, and *R2* is the resistance of the light bulb. Therefore, in our example, the voltage is 9 volts, and the resistance is 1000 and 180 ohms:

$I = V/(R1 + R2)$ = 9 volts/(1000 + 180 ohms) = 0.076 amps

Notice that the current in this experiment is less than the current in Experiments 1 and 2. This is because the current encounters more resistance as it flows through the circuit.

EXPERIMENT 4: PARALLEL RESISTORS

Experiment 3 examined resistive loads in series. Experiment 4 examines resistive loads placed in parallel.

Experiment 4 requires the following materials:

- 9-volt battery
- 9-volt battery connector
- Light bulb with wires
- One ¼ watt, 1 K ohm resistor

1. Connect one wire from each of the resistor, the battery connector, and the light bulb.

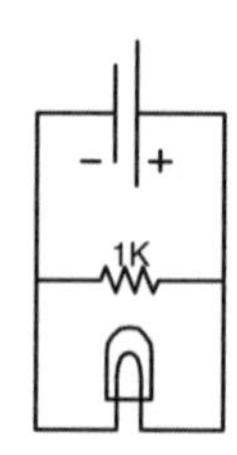

Figure 3.7. A light bulb and resistor in parallel.

2. Connect the remaining wire from each of the resistor, the battery connector, and the light bulb.
3. Attach the 9-volt battery to the battery connector.
4. Observe the light bulb and its brightness. Is it brighter than the light from Experiment 1? Is it brighter than the light from Experiment 3?

This experiment shows what happens when you have two resistive loads in parallel. In this case, Ohm's law is represented by $V = I * (R_{total})$. The value for R_{total} is more complex:

$$R_{total} = \frac{R1 \times R2}{R1 + R2}$$

where *R1* is the resistance of the resistor, and *R2* is the resistance of the light bulb. Therefore, in our example, the voltage is 9 volts, and the resistance is 1000 and 180 ohms:

$$I = V / \left(\frac{R1 \times R2}{R1 + R2}\right) = 9\,\text{volts} / \left(\frac{1000 \times 108}{1000 + 108}\,\text{ohms}\right) = 9\ \text{volts} / 152.5\ \text{ohms} = 0.059\ \text{amps}$$

Notice that if you add the current from Experiment 2 and 3, you will get the current above:

$$I_{parallel} = I_{resistor} + I_{light} = 0.009 \text{ amps} + 0.050 \text{ amps} = 0.059 \text{ amps.}$$

This show that the same voltage is available for each device, so each will draw the same current as if it were the only device in the circuit.

If more resistive devices are wired together in parallel, the new resistance is represented by the equation:

$$\frac{1}{R_{total}} = \frac{1}{R1} + \frac{1}{R2} + \frac{1}{R3} + \ldots$$

ELECTRONICS

Many important components are used to build electronic devices. Some of the more prevalent devices are:

- Diodes
- Light-emitting diodes
- Transistors
- Capacitors
- Potentiometers
- Integrated circuits

These devices are shown in Figure 3.8, and their schematic symbol is shown in Figure 3.9.

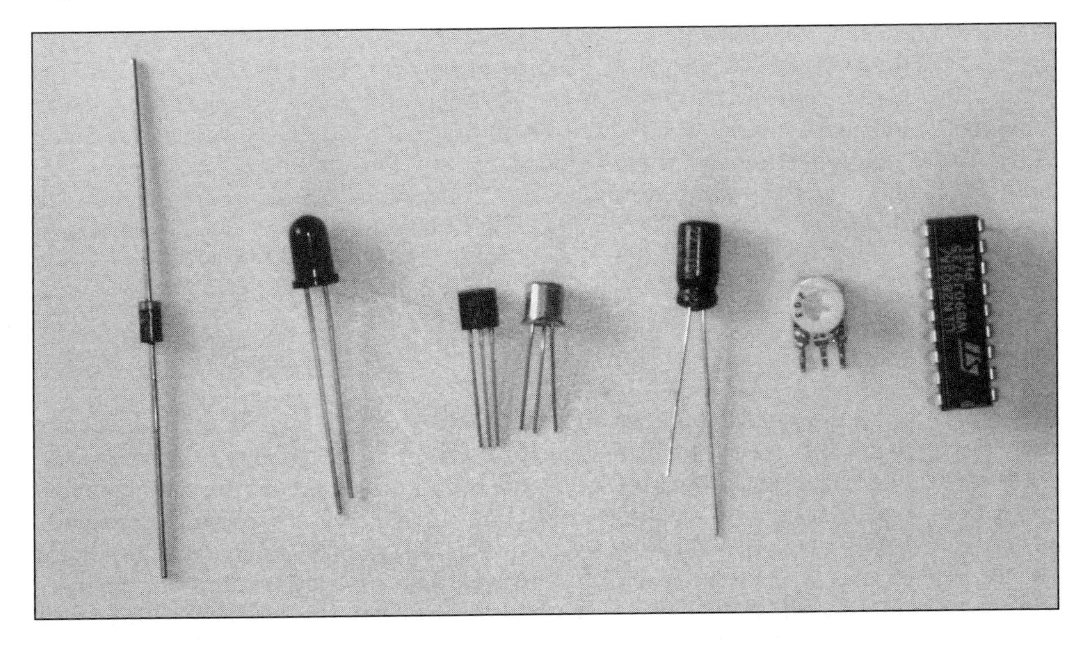

Figure 3.8. Common electronic devices.

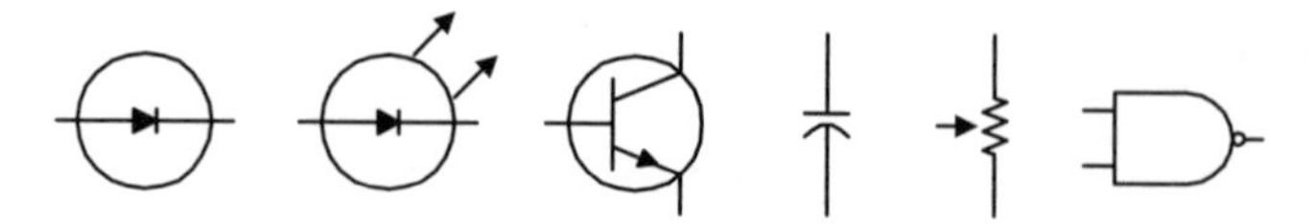

Figure 3.9. Schematic drawings of electronic devices.

Diodes

A diode is a two-electrode semiconductor. It allows current to flow in only one direction. These devices have polarity that must be observed. The negative side, or cathode, has a band around the device. This side points to the negative terminal of a battery.

Light-Emitting Diodes

One special type of diode is a light-emitting diode (LED). This device converts the electrical current into visible light. Unlike the light bulb used in earlier experiments, the LED only heats up or excites a gas inside the device and does not rely on heat for its light. Like a typical diode, it allows current to flow from the anode (positive side) to the cathode. The cathode side is typically notched, and the cathode wire is typically shorter than the anode.

Transistors

A transistor is a semiconductor device that contains three electrodes. It is used as an amplification device, or as an on/off switch. It allows current to flow from one wire of the device to another. One wire, called the *base,* serves as the on/off switch. Another wire, called the *collector,* serves as the source of current. The last wire, the *emitter,* passes this current out of the device if the base has a positive voltage. Transistors come in several shapes and sizes. Figure 3.8 shows two popular sizes. The transistor in the metal container is used when a lot of current flows through the device. The metal pulls the heat generated by the current away from the semiconductor. If the transistor will not switch or amplify much current, then an inexpensive plastic case is used.

Capacitors

A capacitor is a device that opposes any change in circuit voltage. It will hold a voltage charge within the device itself. A capacitor is used in electronics to store charge that may be used later. It also is used to keep a constant voltage across a single integrated circuit even if the source voltage rises or lowers slightly. A capacitor starts with no charge. When a battery or other power source is attached, the capacitor builds up an electrical charge. When the capacitor is charged to its maximum potential, the electric potential is discharged. If the battery remains attached, the rate of discharge will be equal to the rate of charging. The capacity of a capacitor is measured in farads, or more commonly, microfarads.

Potentiometers

A potentiometer is simply a variable resistor device. Typical inexpensive resistors have a known resistive value. A potentiometer allows the user to change that ohm resistance by turning a wheel or a screw on the device.

Integrated Circuits

An integrated circuit is the basis of all electronic devices. A typical integrated circuit chip is made of plastic and has metal legs. Deep inside this piece of plastic is a small piece of silicon. On the silicon is the circuitry that holds a particular function, which is whatever was manufactured for that specific integrated circuit. The metal legs of the circuit take the electrical current from outside to inside the plastic to the piece of silicon. The functionality on the silicon is exercised, which may include directing current back out of the chip.

Integrated circuit chips have a whole range of functions, ranging from the most complex computer processor devices, like the Pentium® II central processing unit, down to simple functions like a device that will invert an electrical voltage from 0 volts to 5 volts. In the case of the Stiquito walking circuit described in Chapter 8, we use one integrated circuit that has several transistors on it and one that creates a periodic pulse. This walking circuit is one simple example of how integrated circuits can miniaturize the function of several larger, but similar devices.

Experiments 5 through 7 allow you to investigate the electrical properties of the electronics included in the Stiquito Educational Kit.

EXPERIMENT 5: LIGHT-EMITTING DIODES

Experiment 5 demonstrates the properties of an LED. You may need to conduct these experiments in a room with dimmed lights to view the LED light.

Experiment 5 requires the following materials:

- 9-volt battery
- 9-volt battery connector
- One integrated LED

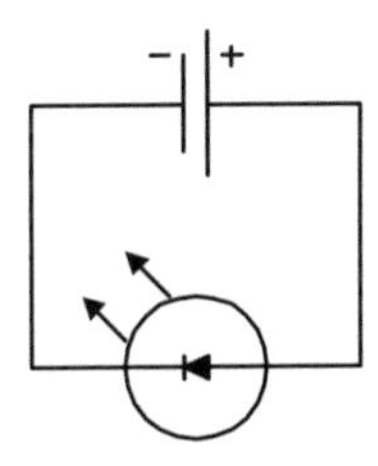

Figure 3.10. LED experiment schematic.

1. Attach the red wire of the battery connector to the longer wire of the LED.
2. Attach the black wire of the battery terminal to the shorter wire of the LED.
3. Attach the battery to the battery connector. Observe that the light shines brightly.
4. Touch the device to see if it is hot.
5. Remove the LED from the battery connector and attach the red wire to the shorter wire of the LED, and the black wire to the longer wire of the LED. Does the diode light?

This experiment shows that the diode only allows current to flow in one direction. Notice that the LED only lights when the cathode side of the LED, the side with the notch, is toward the negative terminal of the battery.

The LED in the Stiquito Educational Kit is actually an integrated LED, which means the plastic lens contains both a small 100 ohm resistor and a diode. The resistor is used to ensure that an infinite amount of current does not flow through the LED.

EXPERIMENT 6: TRANSISTORS

Experiment 6 demonstrates how a transistor can be used as a switch. We will use the transistor to turn an LED on and off.

Experiment 6 requires the following materials:

- 9-volt battery
- 9-volt battery connector
- Wire wrap wire
- One ¼ watt, 1 K ohm resistor
- One integrated LED
- One 2N2222 transistor
- One momentary switch

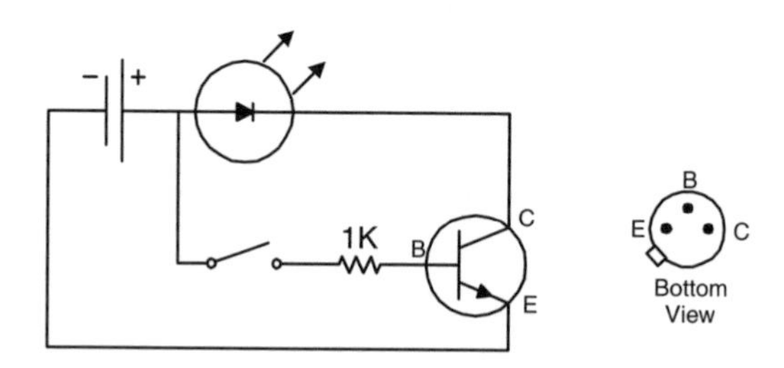

Figure 3.11. Transistor experiment schematic.

Examine the transistor included in your kit. Look at Figure 3.11 and notice how the three wires are labeled C, B, and E for collector, base, and emitter. For the transistor device included in the kit, current flows from the collector to the base to the emitter, based on the presence or absence of a voltage at the base.

1. Using the battery connector, resistor, LED, transistor, and switch, wire the circuit as shown in Figure 3.11. Use the wire wrap wire where needed. Remember to wire the LED with the cathode toward the negative side of the battery.
2. Attach the battery to the battery connector, but do not activate the switch. Do you see the light bulb lighting? When there is no voltage across the base of this transistor, the current is not allowed to flow through the transistor.
3. Now press the momentary switch. When voltage flows across the base of this transistor, the current is allowed to flow through the transistor. Since current flows through the transistor, current also flows through the LED, so it will light.
4. Touch the transistor. Is it hot?

The transistor should not get too hot since there is only a little current flowing through it. Later in Chapters 7 and 8, the transistors used will carry a lot of current. Be careful! Sometimes transistors get so hot they can give you a serious burn.

EXPERIMENT 7: CAPACITORS

Experiment 7 demonstrates how a capacitor stores an electrical charge and then uses it later.

Experiment 7 requires the following materials:

- 9-volt battery
- 9-volt battery connector
- Wire wrap wire
- One ¼ watt, 100K ohm resistor
- One integrated LED
- One 2.2 microfarad, 25 volt capacitor
- One momentary switch

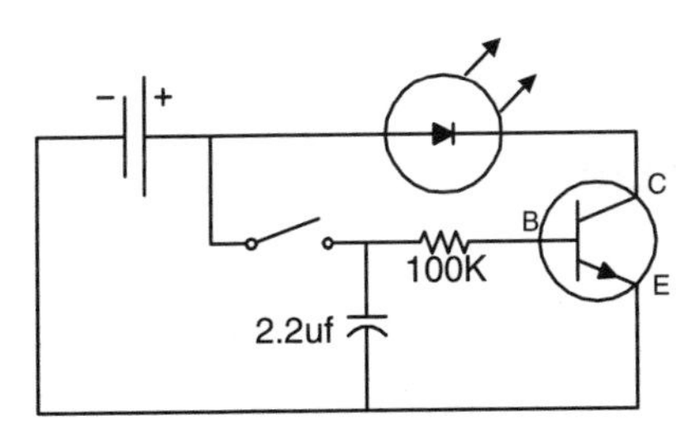

Figure 3.12. Capacitor experiment schematic.

Many capacitors have a distinct polarity. The case near one of the wires will be marked as positive (+) or negative (–). This marked wire should point toward the corresponding battery voltage. For example, if the capacitor wire is marked (–), then that wire should be connected to the negative terminal on the battery (the black wire).

1. Using the battery connector, resistor, LED, transistor, capacitor, and switch, wire the circuit as shown in Figure 3.12. Use the wire wrap wire where needed. Remember to wire the LED with the cathode toward the negative side of the battery. ***For safety reasons, make sure you wire the capacitor correctly!***
2. Attach the battery to the battery connector, but do not activate the switch. Do you see the light bulb lighting? When no voltage is stored in the capacitor, it will not light the LED.
3. Now press the momentary switch. The capacitor will charge. A separate event is that the LED will light. The power for the LED is still coming from the battery. Voltage across the base of this transistor is coming from the battery; the current is allowed to flow through the transistor. Since current flows through the transistor, current also flows through the LED, so it will light.
4. Release the switch. The LED will light for a short time. In this case, the capacitor is providing the current to the base of the transistor, allowing current to flow through the LED and transistor. When the capacitor has discharged all of its stored charge, the voltage at the transistor base will be 0v, and current will no longer flow through the transistor. The LED will no longer light.

The rate of discharge for a capacitor is $t = 1/(R*C)$, where t is the time, R is the resistive load draining the capacitor, and C is the charging capacity of a capacitor in farads. In the example above, the values for R = 100000 ohms and C = 0.0000022 farads. Therefore:

$t = 1/(100000 * 0.0000022) = 4.55$ seconds

You will notice, however, that the LED lights for a shorter time (maybe 1 second). Why? The LED requires 0.012 amps of current to light. As the capacitor is discharged, the voltage and current at the base of the transistor drops so low that it no longer drives at least 0.012 amps of current through the emitter and collector. Although some current is still flowing, there is not enough to light the LED.

Capacitors are used extensively in the electronics industry. Because of their ability to stabilize voltages, they are often paired one-for-one with each integrated circuit on a circuit board.

READING RESISTOR VALUES

Many cylindrical resistors, including carbon-film and metal-film resistors, are marked with colored bands around their circumference. These bands identify the value of the resistance in ohms and accuracy. The rings may be grouped closer to one side of the resistor, but today's resistors usually have evenly spaced bands across them. See Figure 3.13 for an example of two resistors.

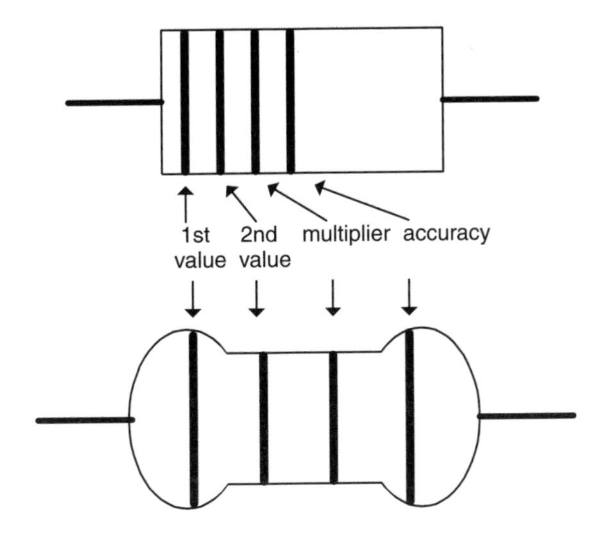

Figure 3.13. Reading the bands of a resistor.

The colors of the bands are important. The first two colored bands on the left side represent the value. This value is then multiplied by a factor of 10, as identified by the color of the third band. The fourth band identifies the accuracy, or tolerance, of the resistor. The colors and their value are shown in Table 3.2.

Table 3.2. Resistor Color Codes

Color	First Value	Second Value	Multiplier	Accuracy (Tolerance)
none	–	–	–	+/–20%
Silver	–	–	× 0.01	+/–10%
Gold	–	–	× 0.1	+/–5%
Black	0	0	× 1	–
Brown	1	1	× 10	–
Red	2	2	× 100	–
Orange	3	3	× 1,000	–
Yellow	4	4	× 10,000	–
Green	5	5	× 100,000	–
Blue	6	6	× 1,000,000	–
Violet	7	7	× 10,000,000	–
Gray	8	8	× 100,000,000	–
White	9	9	–	–

For example, if we have a resistor with the colors Brown-Black-Orange-Gold as the first, second, third, and fourth rings, the resistor value will be:

First value	Second value	Multiplier	Accuracy	Value
Brown	Black	Orange	Gold	
1	0	× 1,000	+/–5%	= 10 × 1,000 = 10,000 = 10k ohm

MORE ELECTRONICS TO COME!

Chapters 7 and 8 include more examples of electronic circuits; they give you the opportunity to build working electrical circuits that control the Stiquito robot. You can also use the electronics described in this chapter to perform more experiments on your own.

EXERCISES

1. Examine the circuit in Experiment 2. Calculate the current if a 470 ohm resistor is used instead.
2. Examine the circuit in Experiment 3. Calculate the current if a 470 ohm and a 10K ohm resistor are connected in series.
3. Examine the circuit in Experiment 4. Calculate the current if a 470 ohm and a 10K ohm resistor are connected in parallel.
4. Using a battery, a red LED, and a green LED, design a circuit that will light the green LED when a battery is attached correctly. The red LED should light when a battery is attached backwards. Draw the circuit schematic.
5. Examine the circuit in Experiment 7. Modify the circuit so that the LED will stay lit longer.
    a. Draw the circuit schematic.
    b. Wire the circuit.
    c. Run an experiment to determine the time it takes to discharge the capacitor. Record the time for many trials and compute the average.
6. Identify the values of the following resistors, based on color:
    a. Red-Red-Red-Gold
    b. Brown-Green-Orange-Silver
    c. Brown-Black-Brown-None
    d. Blue-Gray-Green-Gold
    e. Gold-Yellow-Brown-White
7. Identify the color bands on the following resistor values:
    a. 5.6 ohm
    b. 4.7k ohm
    c. 330k ohm
    d. 1.2M ohm

BIBLIOGRAPHY

Bonnet, B., and D. Keen. 1996. *Science fair projects with electricity and electronics.* New York: Sterling Publishing.

Horn, D. T. 1994. *Basic electronics theory with projects and experiments.* Ridge Summit, Penn.: TAB Books.

Leon, G. 1991. *Electronics projects for young scientists.* New York: FranklinWatts Publishing.

Miller, R. 1988. *Electronics the easy way.* New York: Barron's Educational Series.

Schuler, C. A., and R. J. Fowler. 1988. *Experiments in basic electricity and electronics.* New York: McGraw-Hill.

Chapter 4

Nitinol Basics

James M. Conrad and Wayne Brown

When it was first announced that a "robot muscle" had been invented and that it was usable, strong, and lightweight, the resounding cry was "boy, aren't we gonna see some cool robots now."

—Mark Tilden, Roboticist

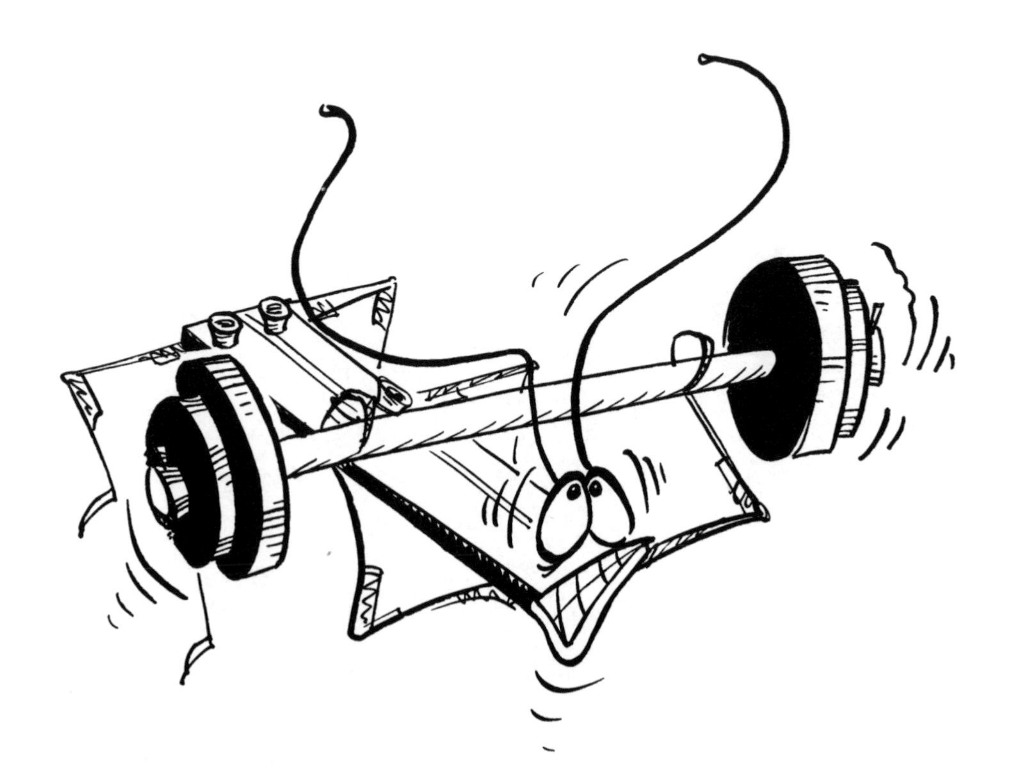

INTRODUCTION

The subject of this book is a small, inexpensive, lightweight, insectoid robot. Small robots similar to Stiquito typically do not use pneumatic air or wheels. Sometimes they use very small motors. Imagine what a robot insect looks like. What you might have envisioned was a slim, small walking insect, using muscles to crawl around.

True muscle-like propulsion did not exist until recently. A new material, nitinol, emulates the operation of a muscle. Nitinol has the properties of contracting when heated, and returning to its original size when cooled. Some opposable force does need to stretch the nitinol back to its original size. This new material has spawned a plethora of new small walking robots that could not be built with motors. Several of these include Stiquito and Stiquito II,[1,2] Boris,[3] and SCORPIO,[4] all of which are shown in Figure 4.1.

(A)

Figure 4.1. Three Nitinol-based insectoid robots: A) Stiquito.

(B)

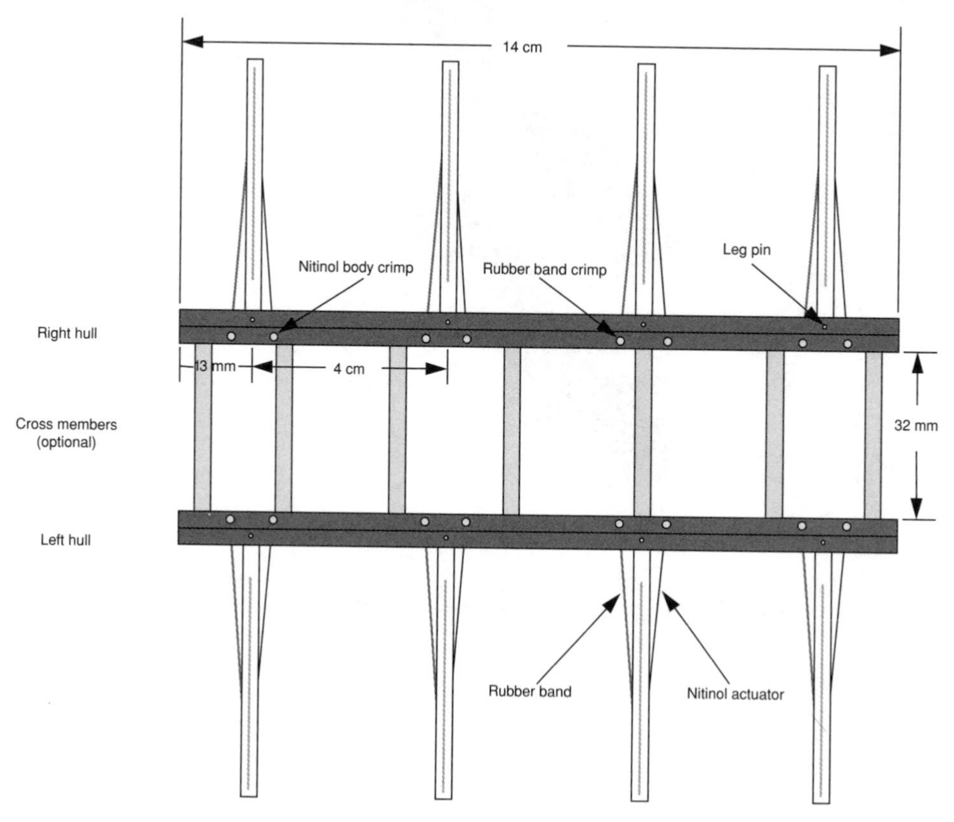

(C)

Figure 4.1. B) Boris, C) SCORPIO.

This chapter discusses using nitinol and includes several experiments on how to use nitinol. Understanding nitinol and how it works will help you when you build Stiquito or other insectoid robots.

NITINOL AND FLEXINOL™ ACTUATOR WIRE

Nitinol is a shape-memory alloy actuator wire made of nickel and titanium. These small diameter wires contract like muscles when electrically heated. This ability to flex or shorten is characteristic of certain alloys that dynamically change their internal structure at certain temperatures. A counterforce, or bias, is required to return the nitinol to its original length or shape.

The nitinol wire translates the heat induced by an electrical current into mechanical motion. The idea of elevating temperature electrically is used in a light bulb. Instead of producing light, nitinol wire contracts by several percent of its length when heated and can then be easily stretched out as it cools. Like the filament in a light bulb, nitinol reacts quickly to both heating and cooling. The contraction of nitinol wires when heated is opposite to ordinary thermal expansion, is larger by a hundred-fold, and exerts tremendous force for its small size. The underlying technology is depicted in Figure 4.2. The main point is that movement is silent, smooth, and powerful, and occurs through an internal "solid state" restructuring of the material.

The function of the nitinol wire is based on the shape-memory phenomenon that occurs in certain alloys in the nickel-titanium family. When nickel and titanium atoms are present in the alloy in almost exactly a 1/1 ratio, the material forms a crystal structure. This structure is capable of undergoing a change from one crystal form to another (a martensitic transformation) at a temperature determined by the exact composition of the alloy. In the crystal form that exists above the transformation temperature (the austenite), the material is high strength and not easily deformed. It behaves mechanically much like stainless steel. Below the transformation temperature, though, when the other crystal form (the martensite) exists, the alloy can be deformed several percent by a very uncommon deformation mechanism that can be reversed when the material is heated and transforms. The low temperature crystal form of the alloy will undergo the reversible deformation easily, so the "memory" strain can be put into the material at rather low stress levels.

An improved derivative of nitinol is Flexinol™, an alloy actuator wire with enhanced movement and stroke endurance characteristics. The contraction of Flexinol™ wire is measured as a percentage of the length of the wire being used and is determined, in part, by the level of stress one uses to reset the wire, or to stretch it in its low temperature phase. In most applications, the bias force is exerted on the wire constantly, and on each cycle as the wire cools, it is elongated by this force. If no force is exerted as the wire cools, very little deformation or stretch occurs in the cool, room temperature state, and correspondingly very little contraction occurs upon heating. Up to a point, the higher the load the longer the stroke. The strength of the wire, its pulling force, and the bias force needed to stretch the wire back out are a function of the wire size or cross sectional area. A pictorial view of these transformations is shown in Figure 4.2.

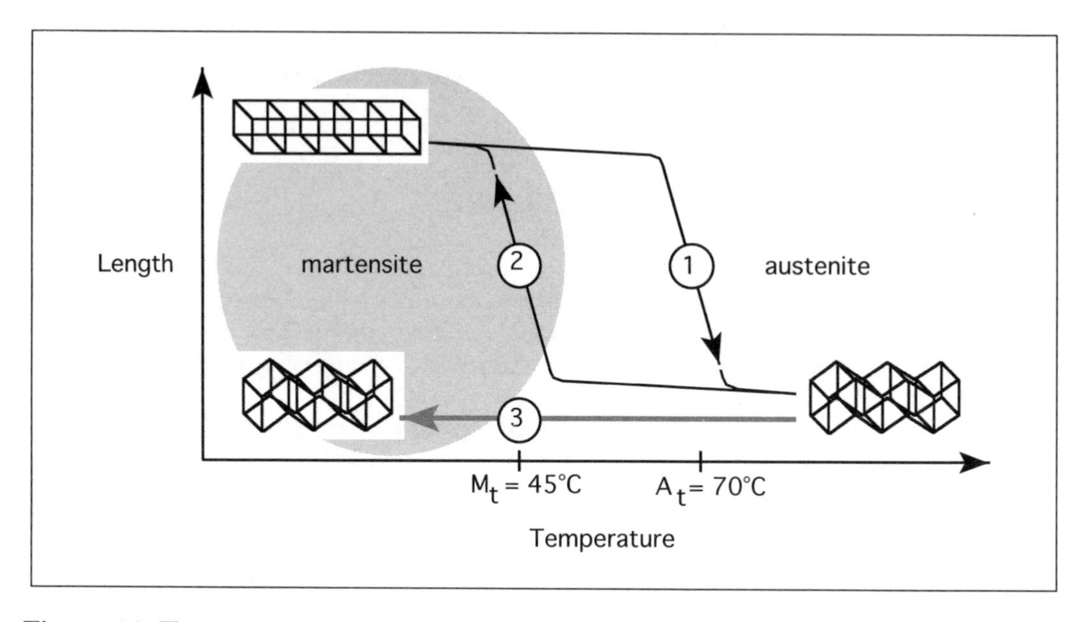

Figure 4.2. The operating cycle of Flexinol™: 1) heating, 2) cooling with a bias force, 3) cooling without a bias force.

Far more important to stroke is how the wire is physically attached and made to operate. Dynamics in applied stress and leverage also vary how much the actuator wires move. While normal bias springs that increase their force as the Flexinol™ contracts induce only a 3% to 4% stroke, reverse bias forces that decrease as the actuator wires contract can readily allow the wire to shrink up to 7%. Mechanics of the device in which it is used can convert this small stroke into movements over 100% of the wires' length and at the same time provide a reverse bias force. The stress or force exerted by Flexinol™ wires is sufficient to be leveraged into significant movement and still be quite strong. Some basic structures, their percent of movement, and the approximate available force they offer in different wire sizes are shown in Table 4.1.

Table 4.1. Stroke and Available Force of Flexinol™

	Approximate Stroke	0.004″ Wire	0.006″ Wire	0.008″ Wire	0.010″ Wire
Dead Weight Bias	4%	150g	330g	590g	930g
Leaf Spring Bias	7%	150g	330g	590g	930g
Simple Lever (6:1 ratio)	30%	22g	47g	84g	133g

ELECTRICAL GUIDELINES

If Flexinol™ wire is used within the guidelines presented in Table 4.2, then tens of millions of cycles can be obtained from the wire. If higher stresses or strains are imposed, then the memory strain is likely to slowly decrease and good motion may be obtained for only hundreds or a few thousands of cycles. The permanent deformation that occurs in the wire during cycling is heavily a function of the stress imposed and the temperature under which the actuator wire is operating. Flexinol™ has been specially processed to minimize this straining, but if the stress is too great or the temperature too high, some permanent strain will occur. Since temperature is directly related to current density passing through the wire care should be taken to heat, but not overheat, the actuator wire. Table 4.2 presents rough guidelines as to how much current and force to expect with various wire sizes.

Table 4.2. Electrical Characteristics of Flexinol™

Wire Diameter Size	Resistance in Ohms per inch	Maximum Pull/Force	Approximate Current at Room Temperature	Contraction Time	Off Time 70° C Wire	Off Time 90° C Wire
0.004″	3	150g	180mA	1 sec.	0.8 sec.	0.4 sec.
0.006″	1.3	330g	400mA	1 sec.	2.0 sec.	1.2 sec.
0.008″	0.8	590g	610mA	1 sec.	3.5 sec.	2.2 sec.
0.010″	0.5	930g	1000mA	1 sec.	5.5 sec.	3.5 sec.

Contraction time is directly related to current input. The figures used here are only approximate since room temperatures, air currents, and heat sinking of specific devices vary. Currents that take approximately 1 second to heat the wire past the austentic transformation temperature, A_t, can be left on without overheating it.

CYCLE TIME

The Flexinol™ wire contracts solely because of the heating and relaxes solely because of the cooling. The wire contracts and relaxes virtually instantaneously because of the temperature of the wire. Consequently, mechanical cycle speed is dependent on and directly related to temperature changes. Applying high currents for short periods of time can quickly heat the wire. It can be heated so fast that the limiting factor is not the rate at which heating can occur but rather the stress created by such rapid movement. If the wire is made to contract too fast with a load, the inertia of the load overstresses the wire. To perform high-speed contractions, inertia must be held low and the current applied in short high bursts. Naturally, current that will heat the wire from room temperature to over 100° C in 1 millisecond will also heat it much hotter if left on for any length of time.

While each device has quite different heat sinking and heating requirements, a simple visual observation can be used to prevent overheating. Measuring the actual internal temperature of the wire across such short time periods is difficult. However, one can tell if the actuator wire is overheated simply by observing if the wire immediately begins to cool and relax when the current is shut off. If it does not begin to relax and elongate under a small load when the power is cut, then the wire has been needlessly overheated and could easily be damaged.

Flexinol™ wire has a high resistance compared to copper and other conductive materials but is still conductive enough to carry current easily. One can immerse the wire in regular tap water and enough current will readily flow through it to heat it. All of the conventional rules for electrical heating apply to the wire, except that its resistance goes down as it is heated through its transformation temperature and contracts. This is contrary to the general rule of increased resistance with increased temperature. Part of this drop in resistance is due to the shortened wire, and part is because the wire gets thicker as it shortens, roughly maintaining its same three-dimensional volume. It makes no difference to the wire whether alternating current, direct current, or pulse width modulated current is used.

Again, relaxation time is the same as cooling time. Cooling is greatly affected by heat sinking and design features. The simplest way to improve the speed of cooling is to use smaller diameter wire. The smaller the diameter the more surface to mass the wire has and the faster it can cool. Additional wire, even multiple strands in parallel, can be used to exert whatever force is needed. The next factor in improving the relaxation or cooling time is to use higher temperature wire. This wire contracts and relaxes at higher temperatures. Accordingly the temperature differential between ambient or room temperature and the wire temperature is greater, and correspondingly the wire will drop below the transition temperature faster in response to the faster rate of heat loss.

Other methods of improved cooling are to use forced air, heat sinks, increased stress (this raises the transition temperature and effectively makes the alloy into a higher transition temperature wire), and liquid coolants.

Miscellaneous

Cutting—Flexinol™ wire is a very hard, anti-corrosive material. It is so hard that cutting it with cutters designed to cut copper and soft electrical conductors will damage the cutters. If you plan to do much work with Flexinol™ wires, purchase a good high quality pair of cutters like those used to cut stainless steel wires.

Attaching—Attaching Flexinol™ wire to make both a physical and an electrical connection can be done in several ways. It can be attached with screws, wedged onto a PC board, glued into a channel with conductive epoxies, and even tied with a knot. The simplest and best way is usually by crimping or splicing. With crimping machines, both electrical wires and hooks or other physical attachments can be joined at the same time. Flexinol™ wire is a very strong material and is not damaged by the crimping process.

Flexinol™ wires tend to maintain the same volume, so when they contract along their length, they simultaneously grow in diameter. This means the wires expand inside the crimps and hold more firmly as the stress increases through pulling. While this works to the advantage in crimps, it can be a disadvantage if glue or solder is used, as the material tends to work itself loose in those cases.

Connecting Materials—When Flexinol™ is heated, its temperature is often over 100° C. The wires often apply pressure with a high force over a small area of the device to which they are attached. It is a good idea to use temperature-resistant materials when connecting them. Such materials, if used in direct contact with the wire, will also need to be nonconductive so it does not provide an electrical path around the Flexinol™ wire. Silicone rubber, the material used to make flexible circuit boards, ceramics, and glass are good examples.

Reverse Bias—Flexinol™ wire contracts about 4.5% when lifting a weight or working against a constant force. This load is also the bias force, which will return the wire to its original length when the wire cools. The length of the stroke can be improved by introducing mechanisms that have a reverse bias force. The bias force is the force that elongates the wire in its rubber-like martensitic phase, returning it to its original shape. A reverse bias force is one that gets weaker as the stroke gets longer (i.e., does not remain constant as in the weight lifting or constant force experiment, which follows). This can be done with leaf springs or with designs that give the nitinol wire a better mechanical advantage over the bias spring, or force, as the stroke progresses.

EXPERIMENTING WITH FLEXINOL™

An important part of building your robot includes understanding how Flexinol™ works. Since the best way to learn is by doing, we have devised three experiments to help you learn about Flexinol™. The experiments consist of three different mechanical setups, but all use the same electrical schematic shown in Figure 4.3.

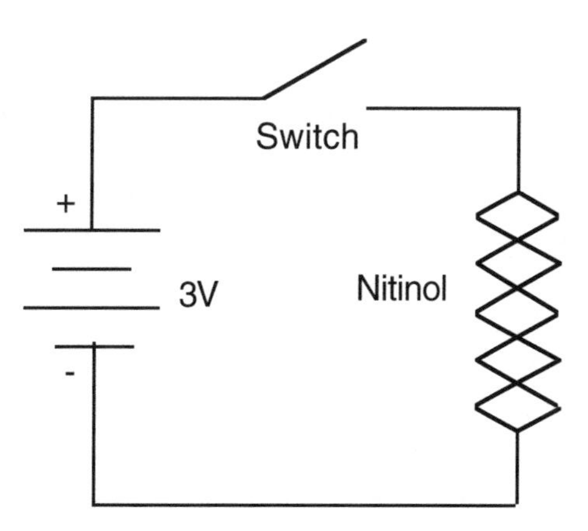

Figure 4.3. The schematic for all experiments.

You will need the following materials to perform these experiments. Keep in mind that each experiment uses a subset of these materials. The following materials are contained in the Stiquito Educational Kit (not included in this book):

- One 4 inch (about 100 mm) length of 0.004 inch diameter Flexinol™ wire, pre-crimped, with insulated wires attached (also available from Dynalloy, Inc.). You may want to buy several and keep them on hand.
- Wire wrap wire, as needed
- One 4 inch (about 100 mm) length of 0.032 inch music wire (spring wire), like K&S Engineering No. 501, also found in hobby stores
- One low voltage momentary switch
- One small plastic bag to hold the coins

You must supply these commonly available materials:

- Two AA batteries
- One 2-AA cell battery holder (available at Radio Shack)
- Five pushpins
- One 2 × 4 × 12 inch wooden board for holding the experiments
- 20 U.S. pennies (1 cent coins) or other 2.5 gram mass
- One 3 by ½ inch piece of wood, cardboard, plastic, or metal to serve as a lever

The only tool you need is a pair of needle-nose pliers.

EXPERIMENT 1: LIFTING DEAD WEIGHT

This first experiment examines lifting a dead weight, 50 grams of coins, and measuring the distance the wire contracts. We specify using 0.004 inch diameter Flexinol™, which has 150 grams of available force. Since one-third of this force is needed to recover the length of the Flexinol™, we use 50 grams of mass. This experiment uses all of the materials except the music wire and lever.

To build the setup for this experiment (see Figure 4.4):

1. Fill the bag with all 20 coins and tie the bag to one crimp grommet of the Flexinol™ wire.
2. Using a pushpin, anchor the other end of the Flexinol™ wire near the top of the 2 × 4 inch board. Allow the bag of coins to hang by the wire. Do not let it touch your tabletop.
3. With a pencil, mark the location of the lower grommet. Later, you will mark the location of this grommet when the Flexinol™ is heated.
4. Insert your two AA batteries into the battery holder, and connect one of the battery wires to one of the insulated wires attached to the Flexinol™.
5. Connect the remaining two wires to the momentary switch.
6. Press the switch ***briefly,*** for no more than 1 second. Observe the contracted length of the wire. Mark the lower grommet hole while the Flexinol™ is heated. Allow the wire to cool fully before you heat it again.

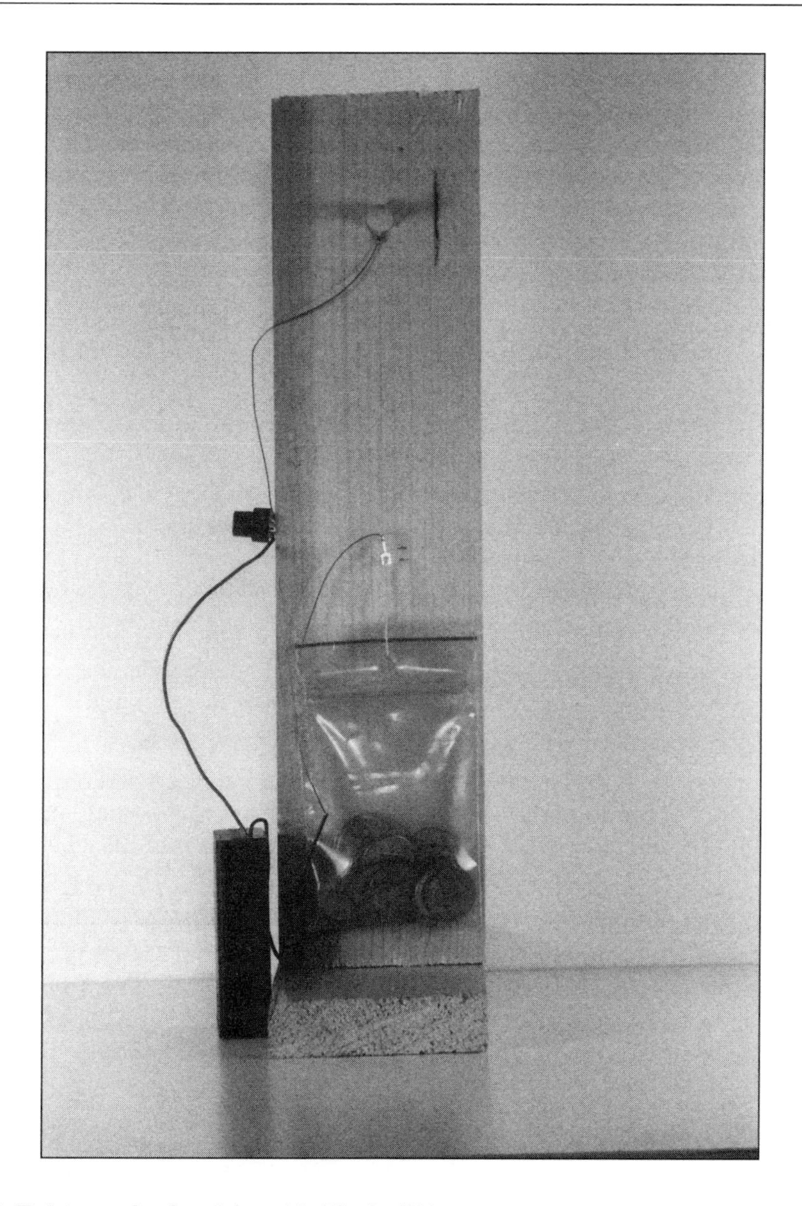

Figure 4.4. Raising a dead weight with Flexinol™.

The result of this experiment is that the wire can lift a lot of mass, but only a short distance. The length the Flexinol™ should contract is 2% to 4%, or only 2 to 4 mm.

EXPERIMENT 2: USING A LEVER

This experiment examines lifting a weight using a mechanical lever. You will measure the distance the lever moves. We specify using 0.004 inch diameter Flexinol™, which has 22 grams of available force for a 6:1 lever. Since one-third of this force is needed to recover the length of the Flexinol™, we use 7.5 grams of mass, equal to three U.S. pennies. This experiment uses all of the materials except the music wire and most of the coins.

To build the setup for this experiment (see Figure 4.5):

1. Make your lever by identifying the pivot point and making a hole big enough for a pushpin. Measure 10 mm from the pivot and make another hole for anchoring the Flexinol™ wire. Measure 50 mm from this second hole and make a third hole for attaching the weight. If you use wood, you may be able to use pushpins instead of making a hole for the Flexinol™ and weight.
2. Place three pennies in the bag and tie the bag to the last hole with wire wrap wire.
3. Attach one crimp grommet of the Flexinol™ wire to the second hole with wire wrap wire.
4. Attach the lever to the 2 × 4 inch board, pushing the pin through the first hole into the board.
5. Using a pushpin, anchor the other end of the Flexinol™ wire near the top of the 2 × 4 inch board. Position the wire so that the lever is 30 degrees below horizontal (at the 4 o'clock position).
6. With a pencil, mark the location of the second and third lever holes. Later, you will mark the location of these holes when the Flexinol™ is heated.
7. Insert your two AA batteries into the battery holder, and connect one of the battery wires to one of the insulated wires attached to the Flexinol™.
8. Connect the remaining two wires to the momentary switch.
9. Press the switch ***briefly,*** for no more than 1 second. Observe the contracted length of the wire and distance the lever rises. Mark the second and third lever holes while the Flexinol™ is heated. Allow the wire to cool fully before you heat it again.

The result of this experiment is that the wire can only lift a smaller mass, but a greater distance than the dead weight. The length the Flexinol™ should contract is 2% to 4%, or only 2 to 4 mm, but the mechanical motion obtained from the lever will be 6 times that, or 12 to 24 mm.

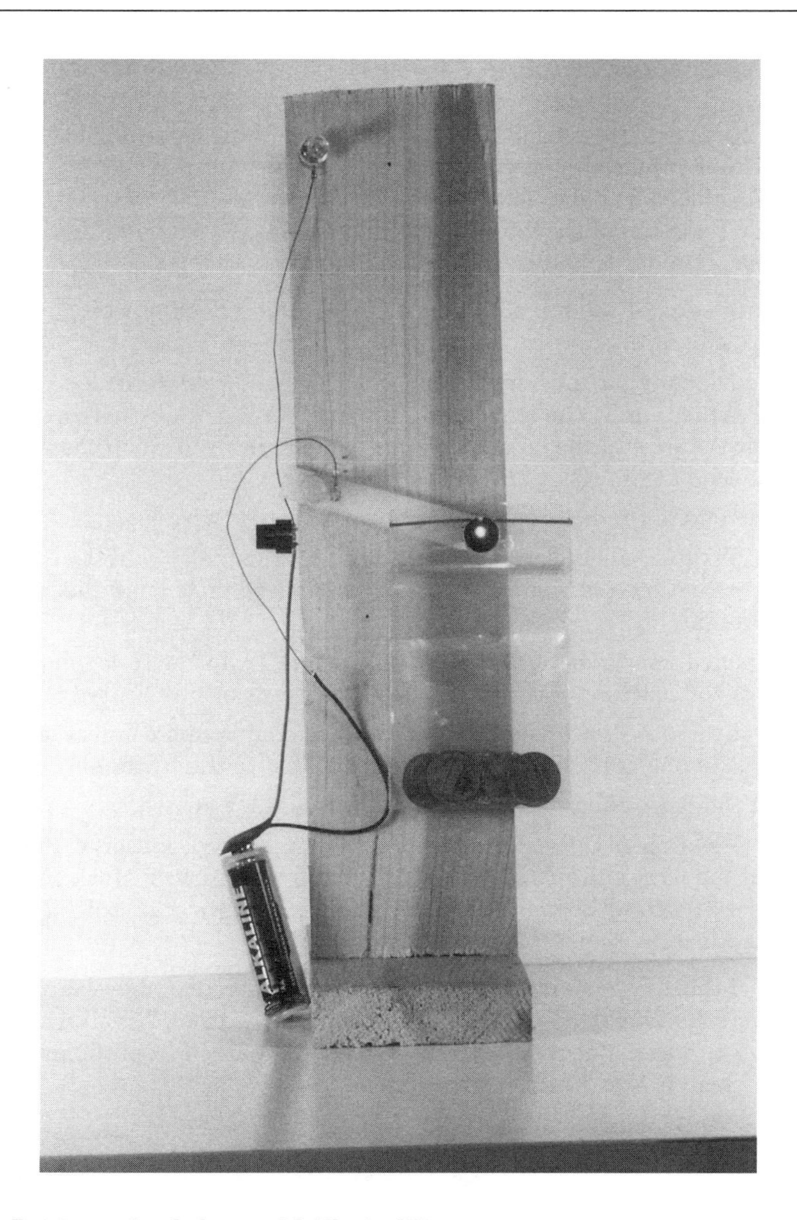

Figure 4.5. Raising a simple lever with Flexinol™.

EXPERIMENT 3: USING A LEAF SPRING

This experiment examines bending a spring wire (Figure 4.6) and allowing the spring to pull the Flexinol™ back into position. You will measure the distance the spring moves. We specify using 0.004 inch diameter Flexinol™, which has 150 grams of available force. This experiment uses all of the materials except the bag, lever, and coins.

To build the setup for this experiment (see Figure 4.7):

1. Bend both ends of the music wire into loops using needle-nose pliers, as shown in Figure 4.6.
2. Using four pushpins, anchor one end of the music wire to the 2 × 4 inch board. Put one of the pins in the loop, and then position the other three at slight angles to anchor the wire to the board. The pin farthest from the loop should be 100 mm from the loop at the other end of the music wire.
3. Hook one end of the Flexinol™ wire onto the music wire loop.
4. Using a pushpin, anchor the other end of the Flexinol™ wire to the 2 × 4 inch board. Position the pin so that the music wire is slightly bent and the Flexinol™ is held taught.
5. With a pencil, mark the location of the Flexinol™ and wire loophole. Later, you will mark the location of this hole when the Flexinol™ is heated.
6. Insert your two AA batteries into the battery holder, and connect one of the battery wires to one of the insulated wires attached to the Flexinol™.
7. Connect the remaining two wires to the momentary switch.
8. Press the switch ***briefly,*** for no more than 1 second. Observe the contracted length of the wire and the distance the spring wire moves. Mark the music wire loophole while the Flexinol™ is heated. Allow the wire to cool fully before you heat it again.

The result of this experiment is that the Flexinol™ wire can move the music wire a greater distance than the dead weight. The length the Flexinol™ should contract is 2% to 4%, or only 2 to 4 mm, but the mechanical motion obtained from the music wire will be 4% to 7%, which is 4 to 7 mm.

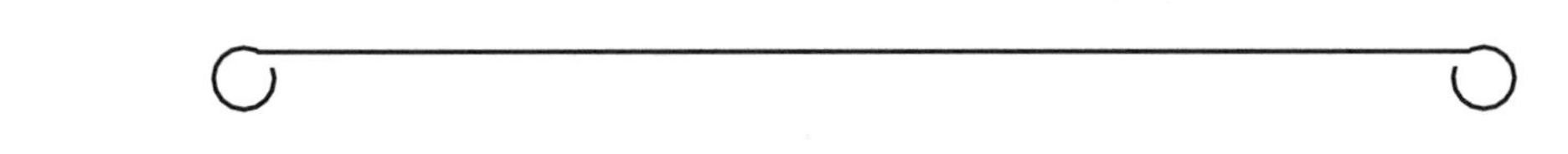

Figure 4.6. Bending the music wire to make a leaf spring.

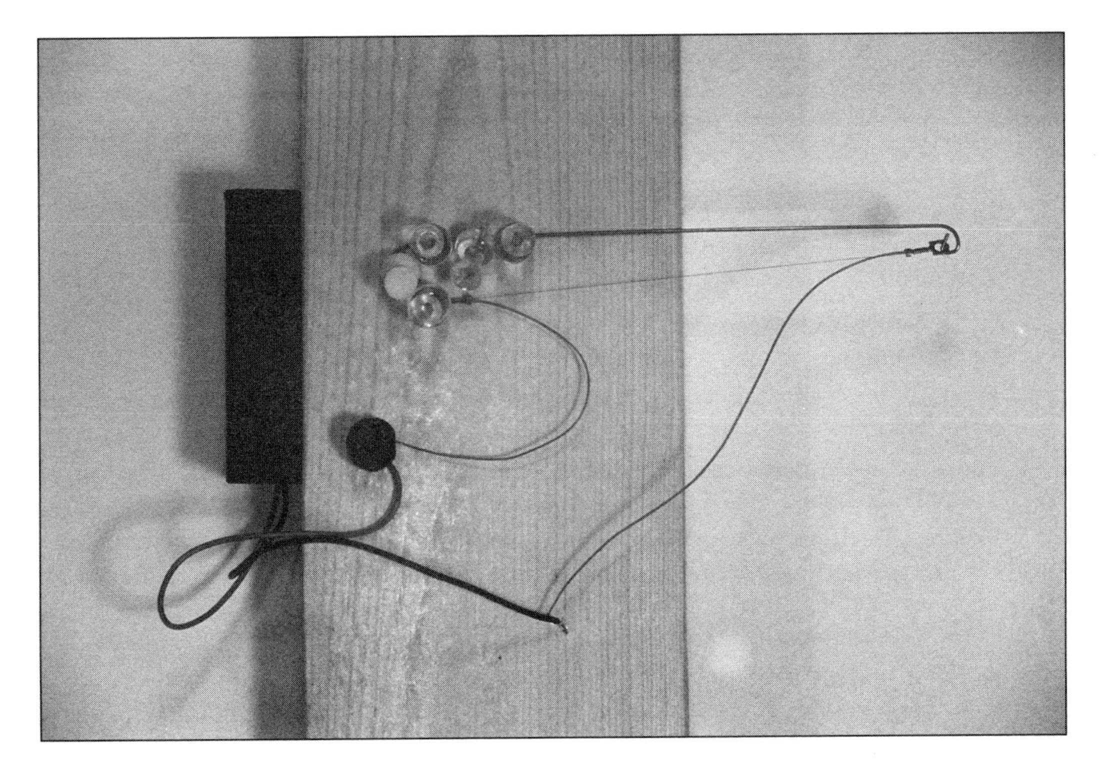

Figure 4.7. Using Flexinol™ and music wire to make a leaf spring bias "leg."

WHERE DO I GO FROM HERE?

These experiments are just a few that can be performed to demonstrate the unique characteristics of nitinol wire. The nitinol in the Stiquito Educational Kit can be used for other experiments you may want to create. If you are interested in other experiments using nitinol, read the *Muscle Wires Project Book* by Roger Gilbertson.[5]

Additional information on nitinol and Flexinol™ can be found at the Dynalloy web site: http://www.dynalloy.com

ACKNOWLEDGMENTS

Much of this article has been taken from "Technical Characteristics of Flexinol™ Actuator Wires," Dynalloy, Inc., Costa Mesa, Calif., 1998.

EXERCISES

1. Investigate the biological workings of a muscle. How does nitinol differ from a muscle?
2. In a certain application, you need to lift 200 kg of weight 5 cm. Based on the approach in Experiment 1, design a device to lift the weight. Use a bundle of 0.004 inch diameter nitinol wires, and assume an unlimited power source. Specifically,
   a. How long will the nitinol bundle need to be?
   b. How many wires do you need in the bundle?
3. Repeat Exercise 2, but use the approach shown in Experiment 2. What will your lever look like?
4. Consider the human body. Experiment 2 roughly emulates the way a major muscle controls a particular human bone. Which muscle and bones operate similarly to Experiment 2?
5. Conduct Experiment 3 and record the time the nitinol takes to cool and expand. Then, conduct Experiment 3 under water, but keep the switch and battery out of the water! Record the time the nitinol takes to cool and expand.
   a. What is the time difference between these two trials?
   b. Substitute water with cooking oil. Now what is the time difference between these two trials?
6. Create an experiment that demonstrates the properties of Flexinol™. The experiment should include:
   a. A prediction (thesis) of what will happen, and why
   b. A materials list
   c. The procedures (steps) to follow
   d. A description of the expected observation
   e. An explanation of the observed behavior.

BIBLIOGRAPHY

1. Mills, J. W. 1992. Stiquito: A small, simple, inexpensive hexapod robot. Technical Report 363a, Computer Science Department. Bloomington, Ind: Indiana University.
2. Conrad, J. M., and J. W. Mills. 1997. *Stiquito: Advanced experiments with a simple and inexpensive hexapod robot.* Los Alamitos, Calif.: IEEE CS Press.
3. Gilbertson, R. G. 1993. *Muscle wires project book.* San Rafael, Calif.: Mondo-Tronics.
4. Conrad and Mills, 1997.
5. Gilbertson, 1993.

Chapter 5

Stiquito: A Small, Simple, Inexpensive Hexapod Robot

Jonathan W. Mills

INTRODUCTION

Legged robots are typically large, complex, and expensive. These factors have limited their use in research and education. Few laboratories can afford to construct 100-legged robot centipedes, or 100 six-legged robots to study emergent cooperative behavior; few universities can give each student in a robotics class his or her own walking robot.

A small, simple, and inexpensive six-legged robot that addresses these needs is described in this chapter. The robot is 75 millimeters long, 70 millimeters wide, 25 millimeters high, and weighs 10 grams. It is constructed of fewer than 40 parts, 12 of which move: Six legs bend in response to six nitinol actuator wires. Nitinol wire, tradenamed Flexinol™, is an alloy of nickel and titanium that contracts when heated. It is also called shape-memory alloy.[1,2] Most parts of the robot perform more than one electrical or mechanical function, but the design can be easily modified. For example, pairs of legs and actuators can be replicated to produce a mechanical centipede with flexible joints between leg segments.

The robot is intended for use as a research and educational platform to study computational sensors,[3,4] subsumption architectures,[5] neural gait control,[6] behavior of social insects,[7] and machine vision.[8] The robot can be powered and controlled through a tether, or autonomously with on-board power supply and electronics.[9] It is capable of carrying up to 50 grams while walking at a speed of 3 to 10 centimeters per minute over slightly textured surfaces such as pressboard, indoor-outdoor carpet, or poured concrete. The feet can be modified to walk on other surfaces. The robot walks when the heat-activated nitinol actuator wires attached to the legs contract (Figure 5.1). Heat is generated by passing an electric current through the nitinol wire. The legs can be actuated individually or in groups to yield tripod, caterpillar, or other gaits. The robot is named Stiquito after its larger and more complex predecessor, Sticky.

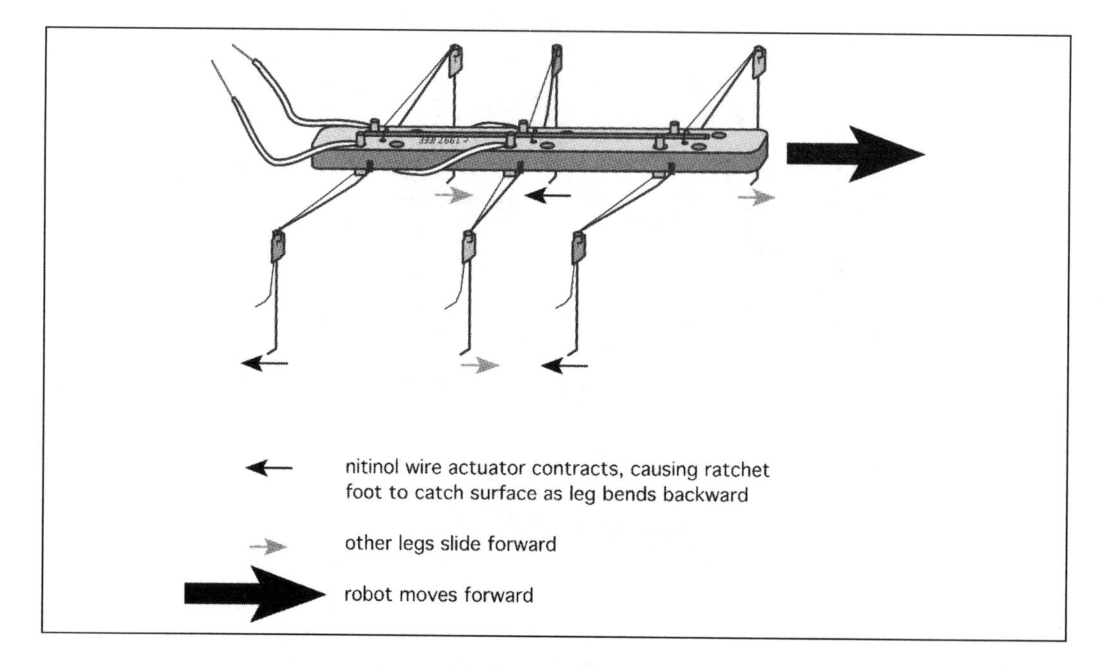

Figure 5.1. How the robot walks.

PREPARING TO BUILD STIQUITO

All materials and tools must be on hand before building the robot. Check the materials in the kit against the parts list. If anything is missing, contact the vendor. The tools needed to construct the robot are typically available in hardware stores, electronics supply stores, or hobby shops. A clear workspace and a relaxed frame of mind will be helpful during construction, especially when installing the nitinol actuators. Correct installation of the actuators will result in a robot that walks well, while a sloppy job will almost certainly lead to one that barely twitches.

Materials List

The robot kit should come with the materials listed here. Please note that some of these materials are used to assemble the manual controller, described in Chapter 6.

Amount	Part number	Item
1 each	ST-100	Molded plastic Stiquito body
1 each	ST-101	Molded plastic manual controller
2 × 100 mm	K&S Engineering No. 100	1/16″ outside diameter aluminum tubing
5 × 100 mm	K&S Engineering No. 499	0.020″ music wire
600 mm	Dynalloy 0.004″ Dia. 70 C	0.004″ (100μm) nitinol wire (Flexinol™)
70 mm	generic	20 AWG copper hook-up wire
600 mm	generic	28 AWG copper wire wrap wire
1,500 mm	generic	34 AWG copper magnet wire
1 each	generic	9-volt terminal assembly
1 each	generic	320-grit sandpaper
1 each	generic	600-grit sandpaper

To complete the robot with manual controller, the following must be purchased:

Amount	Part number	Item
1 each	generic	9-volt battery

Tools List

The following tools are needed to build this robot.

- Needle-nose pliers
- Wire cutters
- Small knife (X-Acto™ type)
- Ruler graded in millimeters
- Voltemeter or two AA batteries and holder

PRECAUTIONS

- Always follow the manufacturer's instructions when using a tool.
- Wear safety glasses to avoid injury to eyes from broken tools, or pieces of plastic or metal that might fly away at high velocity as a result of cutting or sawing. Be especially careful when cutting music wire with wire cutters (cut wire can be forced into a finger, for example).
- Use motorized tools, such as Dremel Moto-tools™, carefully. Motorized tools are not needed to construct Stiquito, although they are helpful if many robots will be built.
- Use a piece of pressboard, dense cardboard, or a cutting board, to protect the work surface, if necessary.

Required Skills

Please remember that assembling Stiquito requires hobby-kit-building skills. This section has been included for the benefit of readers who have not assembled kits before. Practice the skills needed to build Stiquito before assembling the robot, using scrap plastic and thin wire. The kit has extra material in case of mistakes, but there is not enough with which to practice.

Measuring—Following the adage "measure twice, cut once" will prevent most mistakes. Use any metric ruler graded in millimeters. Many figures in this chapter are life-size, so any dimensions included can be used to measure parts.

Cutting—Before cutting, check that your fingers are not in the way of the knife and that a slip of the knife will not damage anything nearby. Direct the knife away from yourself to avoid injury. Make small cuts to avoid removing too much material or making too large or deep a cut.

Deburring—Sawing and cutting can leave rough edges, or burrs, on some parts. Remove the burrs by sanding the rough edge, trimming the burr with a small knife, or lightly abrading the part with a drill bit held in a pin vise. Leaving burrs on parts, especially crimps, can cause the nitinol actuators to break. Parts that are press-fitted can bend or break during assembly if not deburred.

Sanding—The ends of the aluminum tubing should be deburred by sanding them with fine (320-grit) sandpaper. Lightly sand nitinol and music wire with ultrafine (600-grit) sandpaper or emery paper to remove oxide. Sand the wire after it is bent or knotted to avoid breaking it. Sanding wire too much can weaken it enough to break during assembly or when the robot is operating.

Knotting and Crimping Nitinol Wire (Figure 5.2)—Nitinol is similar to stainless steel. The 0.004 inch nitinol wire used in this robot can be knotted without breaking the wire as long as the knot is not tightened excessively. Knotting and crimping nitinol wire is the most reliable way tested to attach the actuators. Nitinol actuators must be taunt, and attached so they cannot pull loose, if this robot is to walk well. The knot-and-crimp attachments have proven reliable for over 300,000 cycles (approximately 100 hours of continuous walking).

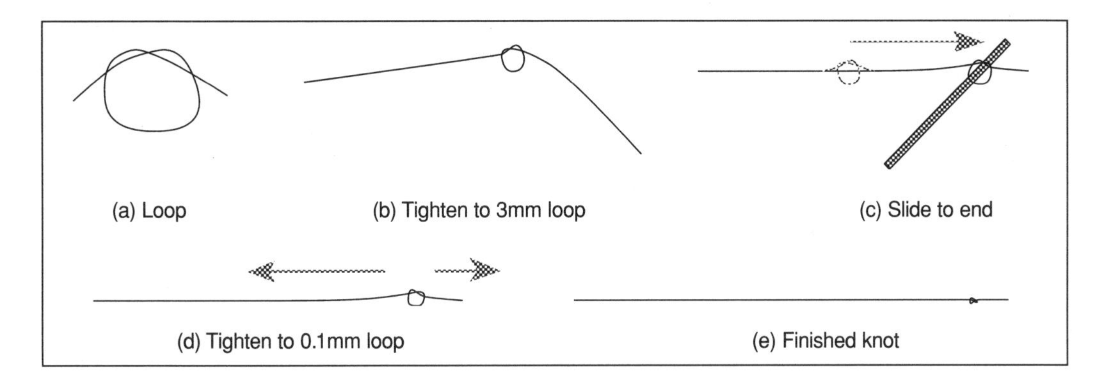

Figure 5.2. Knotting nitinol.

Of the other ways to anchor nitinol actuators, a U-shaped bend in the nitinol wire can pull out of a crimp far enough to reduce leg motion; soldering is difficult to control because the wire contracts and can lose its "memory"; soldering and epoxying nitinol wire might not hold under repeated actuation; and pinning or screwing the nitinol wire to the small parts used in the robot is more complex than knotting and crimping.

To tie a knot, make a loop in the wire, run one end of the wire through the loop to make an overhand knot, then pull by hand to decrease the diameter of the knot's loop to about 3 millimeters. Slide the knot nearly to the end of the wire using a length of stiff wire, then grasp the end of the wire nearest the knot with the needle-nose pliers, and, holding the other end of the wire (or the crimp if one is attached to the other end) with your hand, pull sharply several times with the pliers to tighten the knot. The knot is tight enough when a small loop, approximately 0.1 millimeters in diameter, remains. Tightening the knot further can break the wire.

Crimps are hollow connectors that are squeezed shut to hold, attach, or connect one or more objects (Figure 5.3). This robot uses short lengths of aluminum tubing as crimps to anchor the knotted nitinol actuator wires securely. Two types of crimps are needed. Body crimps hold the actuator wire alone. They are press-fitted into holes in the robot's body to attach the actuator wires indirectly to the control wires; there is no electrical connection between the actuator wire and the plastic body. Leg crimps (Figure 5.4) hold the actuator wire and the music wire leg; they attach directly and electrically connect the actuator wire to the music wire leg.

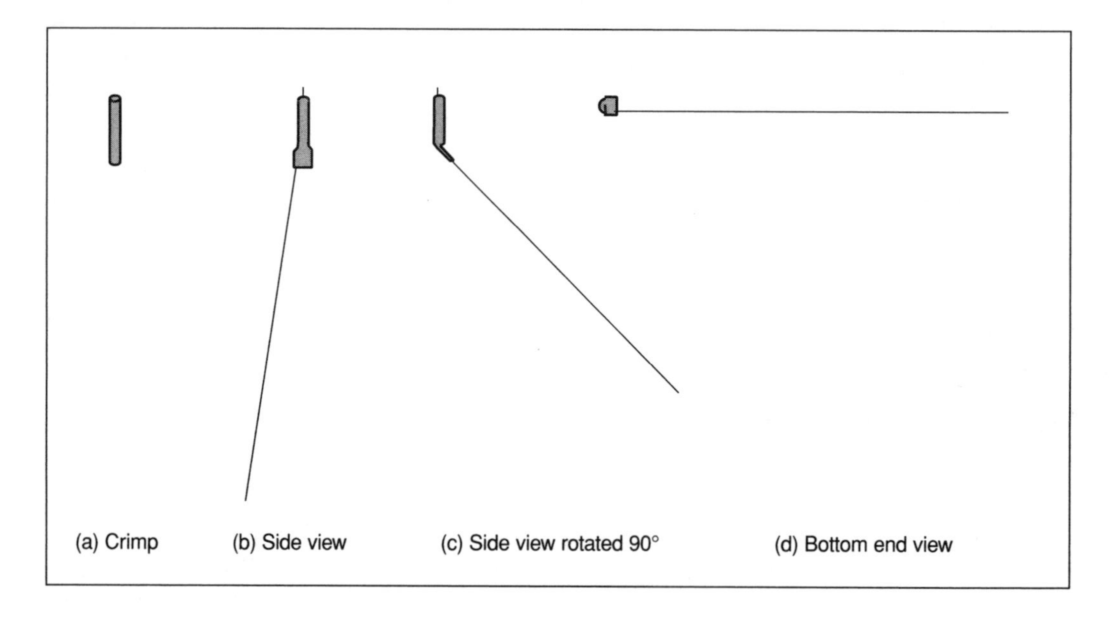

Figure 5.3. Body crimp (views with respect to body crimp).

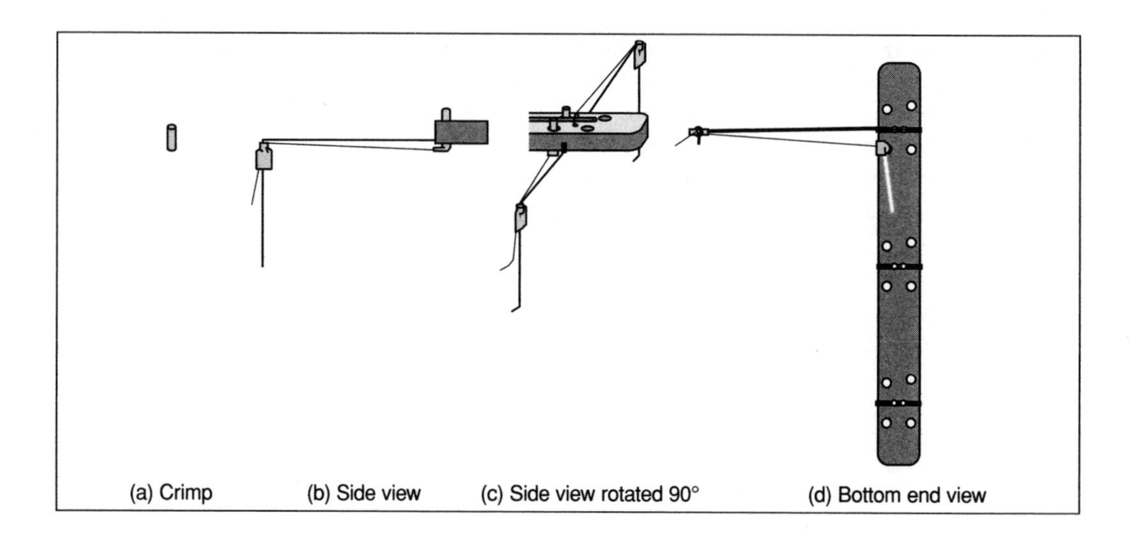

Figure 5.4. Leg crimp (views with respect to leg crimp).

Fixing Mistakes—No matter how carefully one works, mistakes do happen. Most are easy to fix, because almost all steps in the construction of Stiquito allow some tolerance, except for tensioning the nitinol actuator wires, where no slack is allowable. Here are some common problems, with suggestions on how to work around them.

- *Music wire bent incorrectly.* Ninety-degree bends or greater can be rebent gently once or twice before breaking the wire. Bends less than 90 degrees, such as the 15-degree V-clamp in the legs, will usually break if rebent. Stiquito will work with four legs if more music wire cannot be found. Stiquito will work with 0.015-inch-diameter music wire, but cannot carry much weight.
- *Knot in wrong place.* Tie a new knot and keep going. It might be preferable to cut the nitinol in 70 millimeter lengths instead of 60 millimeter lengths for this reason. Untying tight knots in nitinol usually breaks the wire.
- *Crimp must be removed or replaced.* Crimps can be gently squeezed across the wide dimension to undo them, but they should then be discarded. Extra tubing is provided to make new crimps. It will probably be necessary to redo the leg crimps once or twice until you get the hang of tensioning the actuator wires.

CONSTRUCTING STIQUITO

Stiquito has four major assemblies: the body, the legs and power bus, the control wires, and the actuators (Figure 5.5). The actuators are made of nitinol wire.

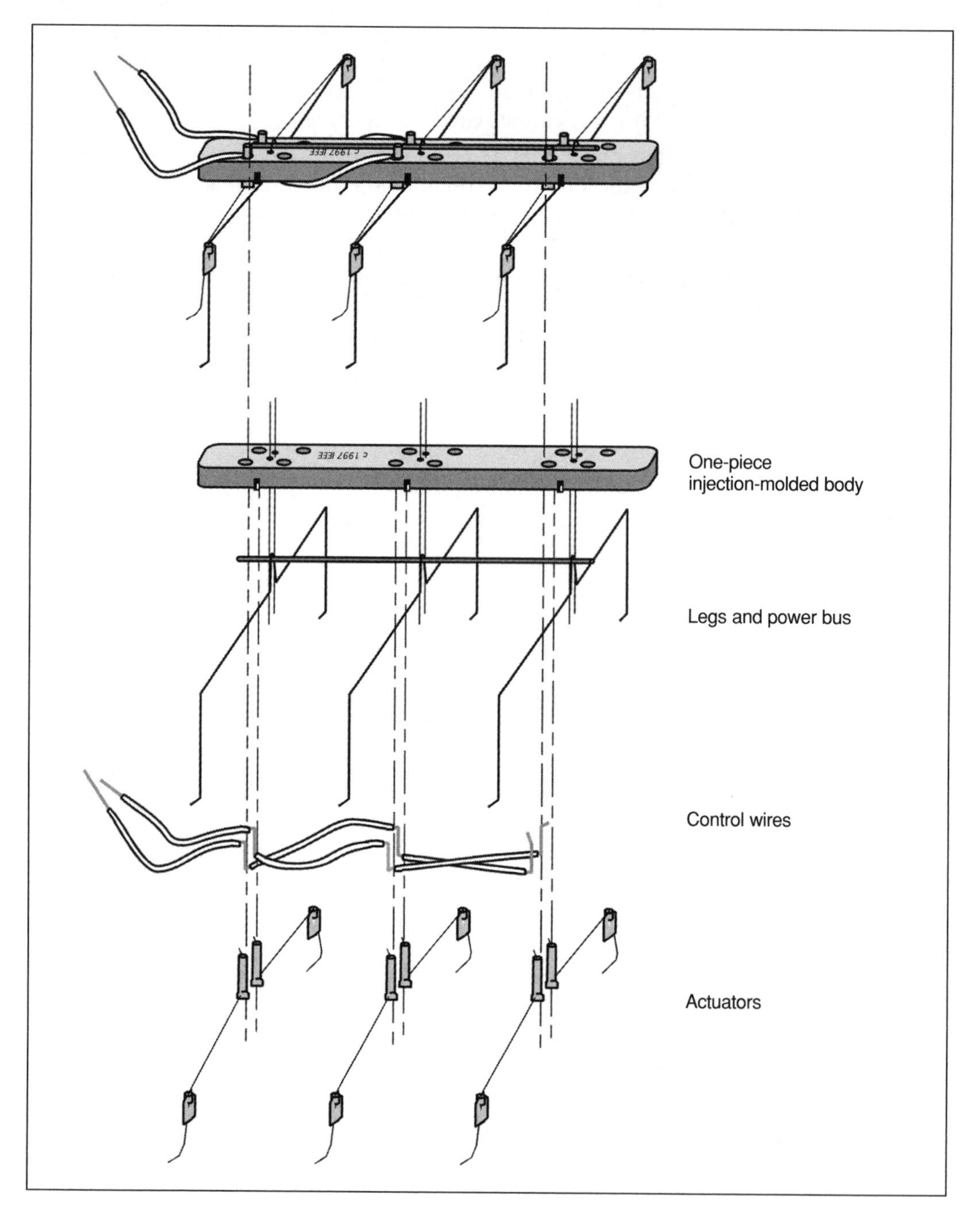

Figure 5.5. Stiquito assemblies.

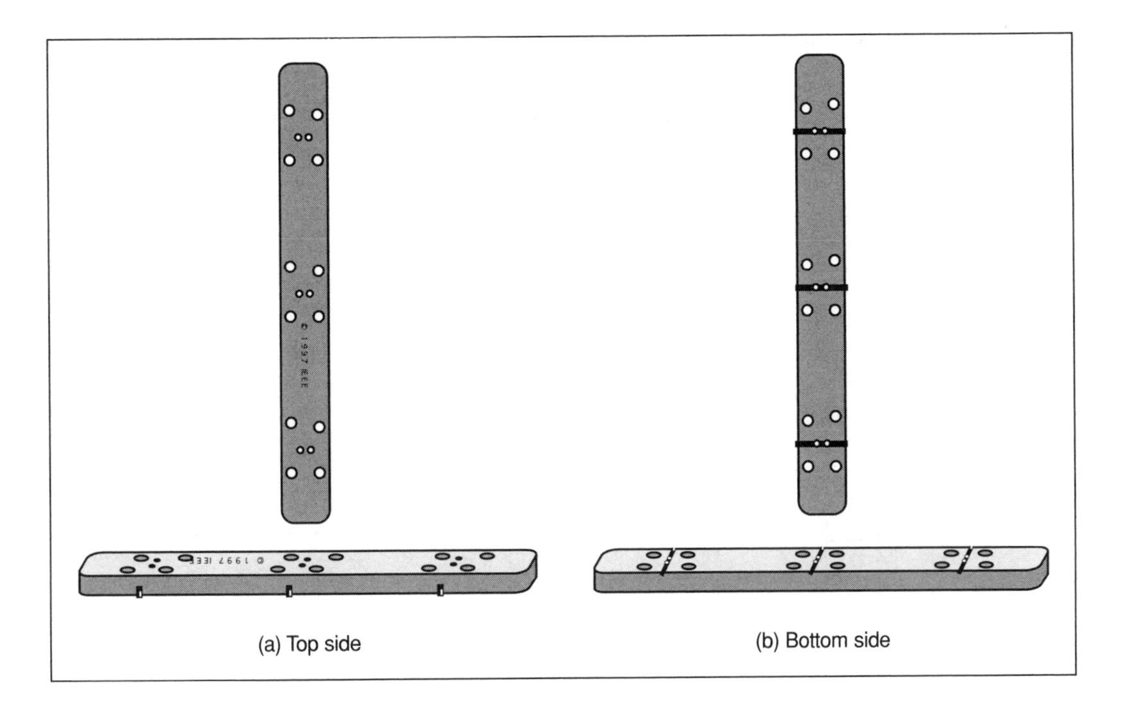

Figure 5.6. The molded body.

The Body

The body (Figure 5.6) provides structural strength and locates the attachment points for the legs and the nitinol actuator wires.

The body is molded with holes and grooves. Examine the plastic to ensure that every hole goes all the way through and that there are no rough edges.

The plastic body has 18 holes. The smallest set of six holes is used to attach the legs to the body. The six large holes that are parallel to the small holes are used to assemble the leg actuators. The following instructions use only these 12 holes. The other set of six holes that have three holes slightly offset can be used for other purposes, including:

- Assembling Stiquito with screws and nuts, as described in Appendix C.
- Assembling Stiquito with legs that have two degrees of freedom, similar to Stiquito II described in Chapter 9.
- Mounting a circuit board for controlling Stiquito, like the circuit described in Chapter 8.

These extra holes add to the flexibility of Stiquito for experimentation.

The Legs and Power Bus

The legs are assembled in pairs from three 100-millimeter lengths of 0.020 inch music wire. The music wire legs perform three functions (See Figure 5.7).

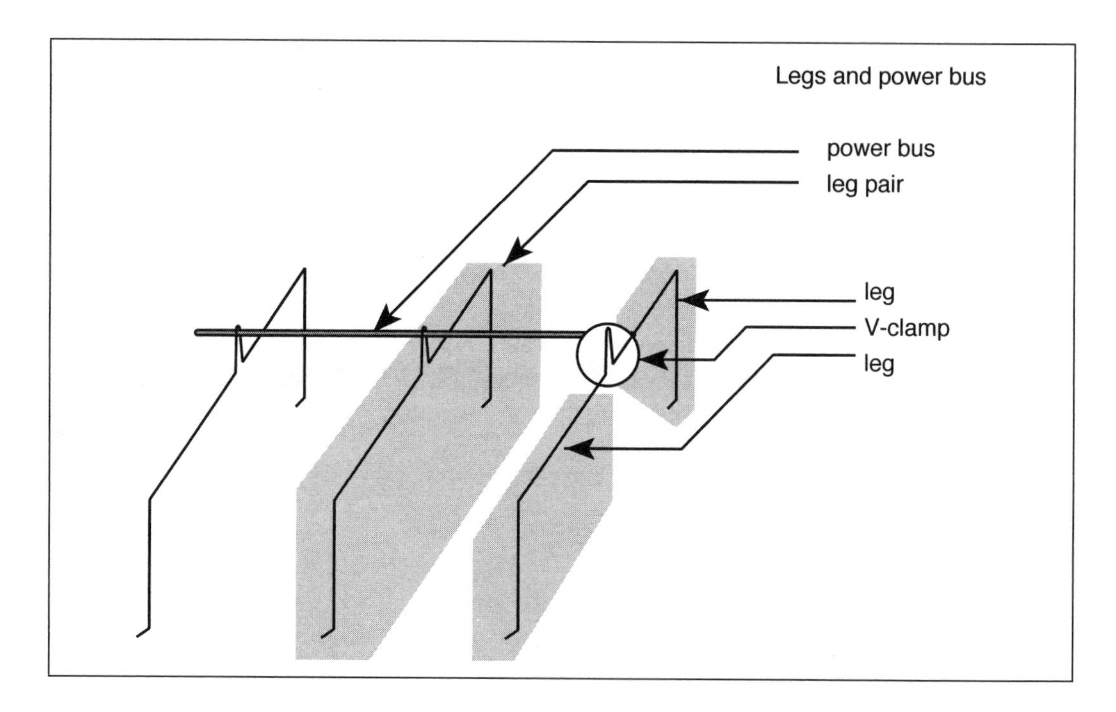

Figure 5.7. Legs and power bus.

1. *Support.* The legs support the weight of Stiquito and its battery and control electronics. Because the wire is bent to fit into a leg clip groove molded in the body, each leg in the pair is mechanically isolated.
2. *Power distribution.* All legs share a common electrical power connection to the power bus and route current to the nitinol actuator wires. The V-bend in the music wire clamps the power bus to the top of the body and electrically connects it to the legs.
3. *Recovery force.* The music wire acts as a leaf spring to provide recovery force for the nitinol wire actuator. Without this spring, or if the actuator is attached loosely, the nitinol wire will contract but will fail to return to its original extended length.

Begin assembling the legs by using three 100-millimeter lengths of 0.020 inch music wire (Figure 5.8). Bend each music wire in the middle to a 15-degree angle. Do not bend the wires too far, or they might crack or break. The apex that forms the V-clamp should be rounded, not sharp. Lightly sand the inside of each V-clamp with the 600-grit sandpaper to remove oxide. (See Figure 5.9.)

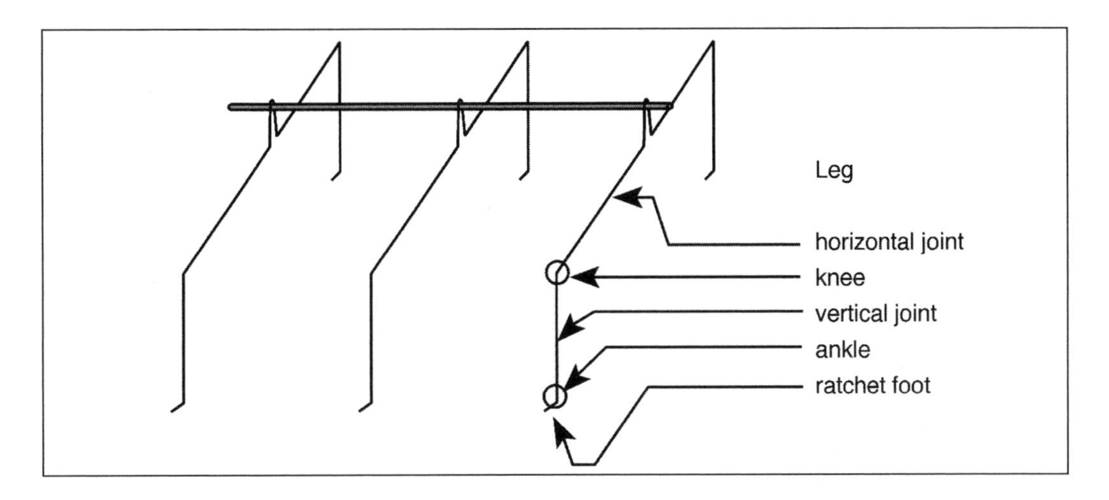

Figure 5.8. Leg detail.

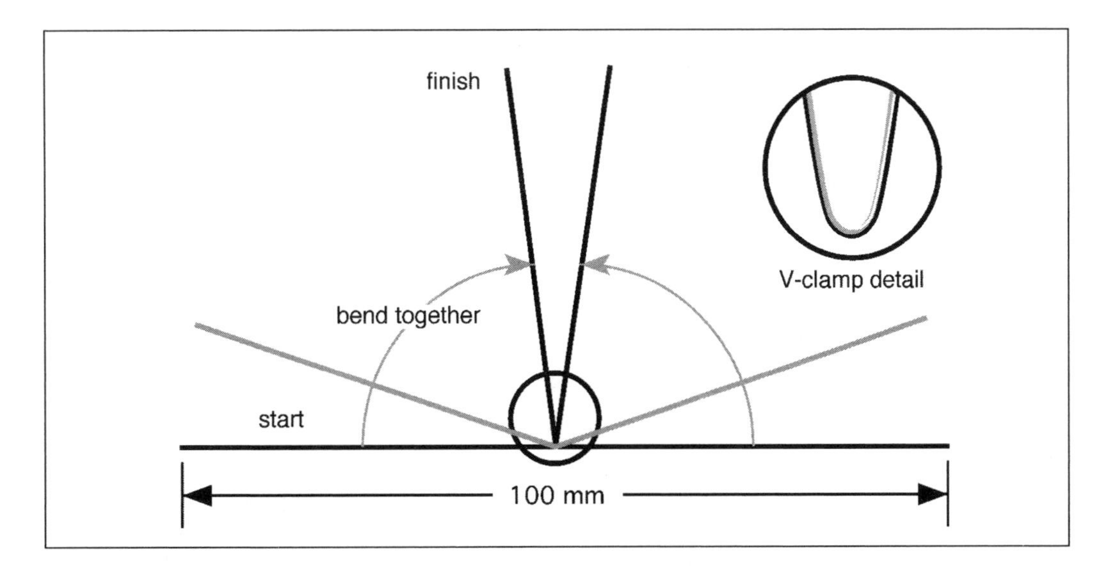

Figure 5.9. Bending the legs (enlargement 6×).

Remove the 70 millimeter length of 20 AWG copper wire from the kit. This is the power bus (Figure 5.10).

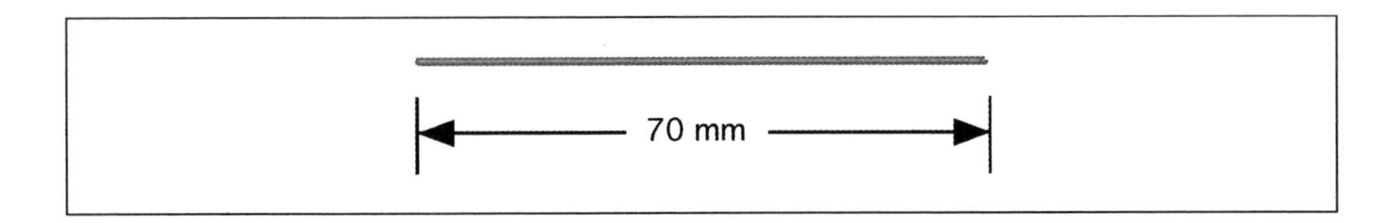

Figure 5.10. Power bus.

Lay the power along the top of the body between the leg holes. Temporarily clamp the power bus by bending the legs together, inserting them through the leg holes, and then pulling the legs through from the other side until the power bus is held tightly by the V-clamp (Figure 5.11).

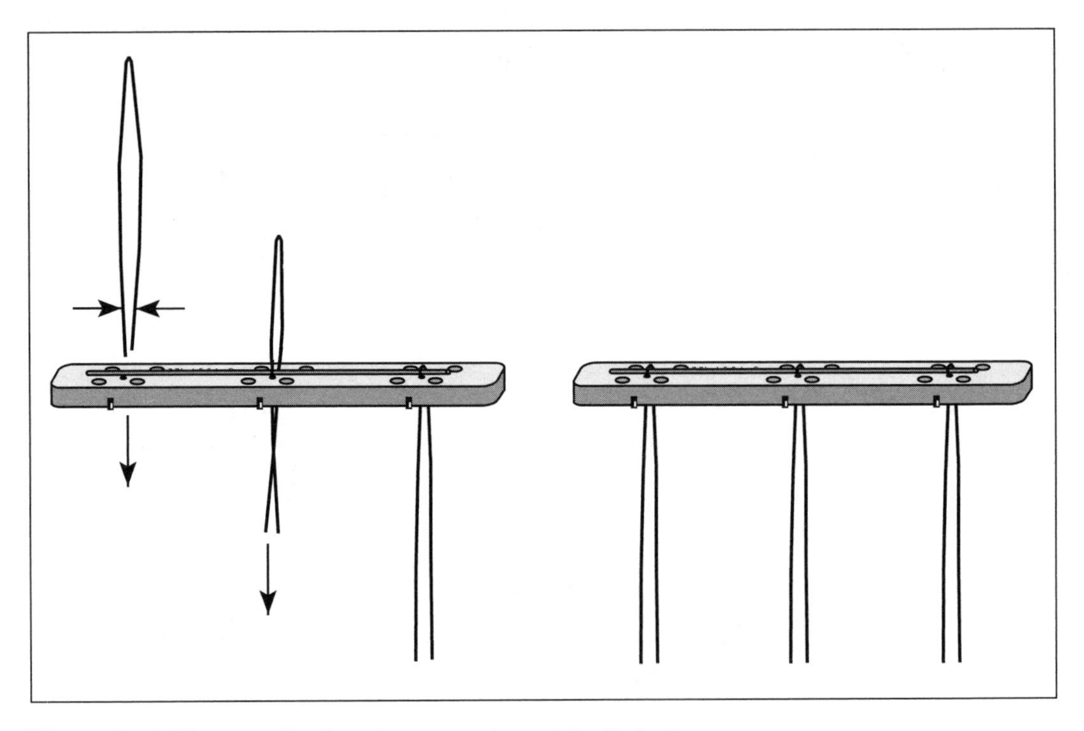

Figure 5.11. Temporarily clamping power bus with all the legs.

Turn the body over and permanently clamp the power bus, simultaneously attaching the legs by spreading each leg in a pair outward by hand, while at the same time pulling upward on the legs. When the legs are almost horizontal, grasp each leg in turn with the needle-nose pliers and firmly bend it downward, while continuing to pull outward, until the leg lies in the clip groove. (See Figure 5.12.)

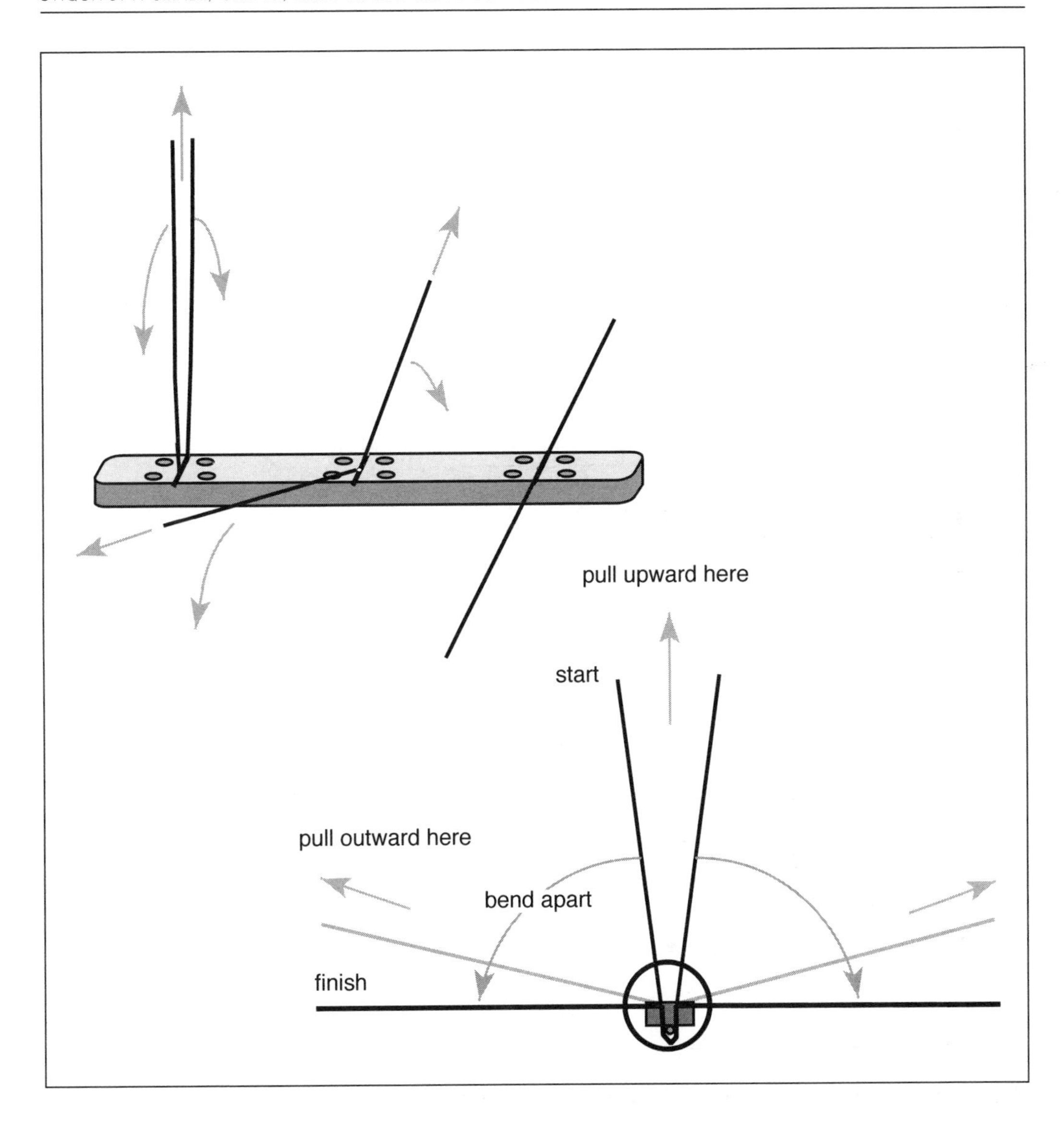

Figure 5.12. Permanently attaching the legs and clamping the power bus.

At this point the power bus should be securely clamped in place. Check to ensure that it will not touch any of the body crimps; insert the length of aluminum tubing and check with a voltmeter. Also check that the electrical connection between the power bus and the legs is good. It should be less than 2 ohms.

Adjust the legs so that they are in a horizontal plane with the bottom of the body and are parallel to each other.

Working with the bottom of the body facing up, form the knee, which separates the horizontal joint from the vertical joint of each leg, by bending the music wire 90 degrees about 30 millimeters from the edge of the body. Adjust the vertical joints so they are parallel (Figure 5.13).

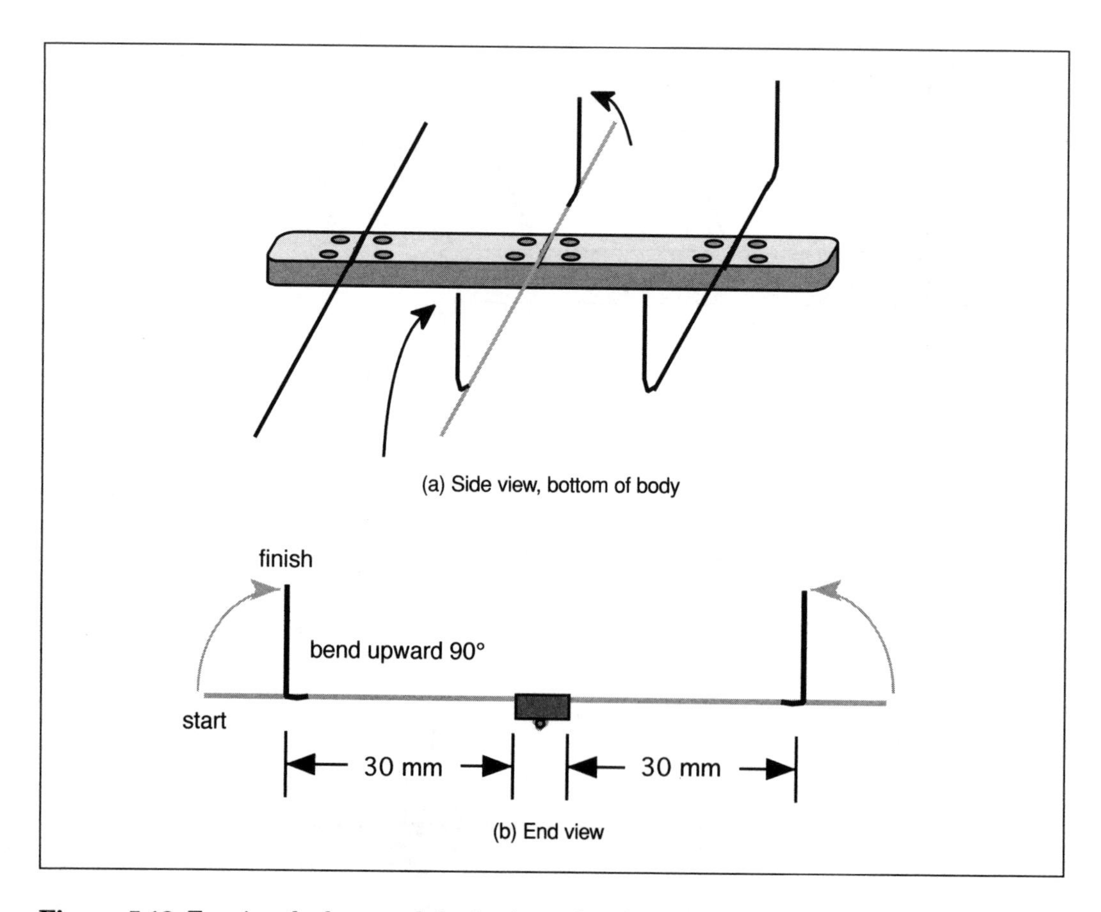

Figure 5.13. Forming the knee and the horizontal and vertical joints.

Trim the vertical joints using the wire cutters so that all legs touch the ground (Figure 5.14). This completes assembly of the legs and power bus.

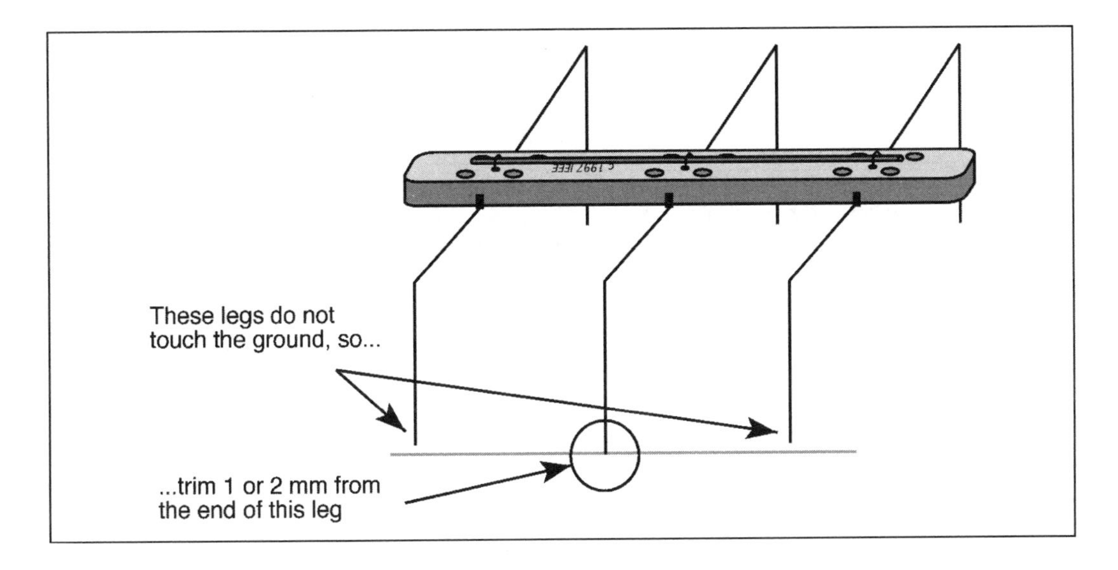

Figure 5.14. Trimming the vertical joints.

Do not bend the music wire to make the ratchet feet now. Wait until the actuators have been completed and tested.

The Control Wires

Cut two 120 millimeter lengths of 28 AWG copper wire wrap wire from the kit for control wires (Figure 5.15). Prepare the control wires for a hardwired tripod gait by stripping 12 millimeters of insulation from each end using the hobby knife.

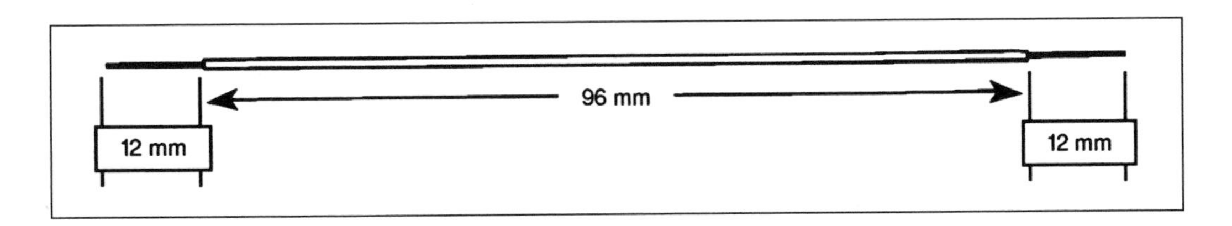

Figure 5.15. Control wire.

Next, separate two lengths of insulation, leaving them on the wire. Use a knife to make a cut in the insulation 36 millimeters from one end of the wire. Make a second cut in the insulation 66 millimeters from the same end. Cut all the way through the insulation, all around the wire. Do not cut the wire. (See Figure 5.16)

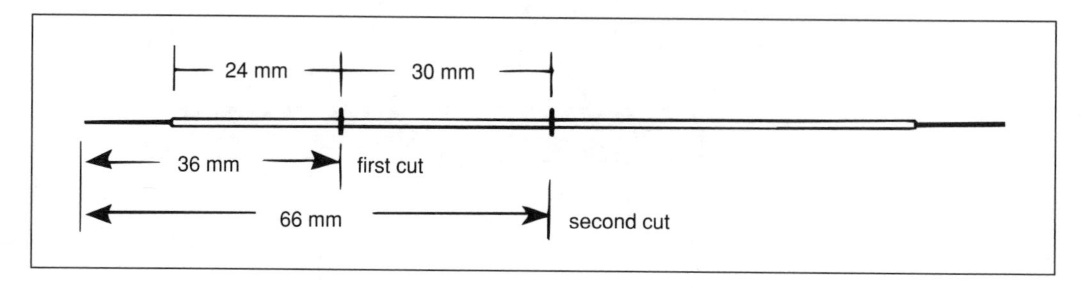

Figure 5.16. Separating two lengths of insulation.

Gently slide the sections of insulation toward the nearest end, leaving 4 millimeters of bare wire at the end and at two places in the middle of the wire. This is a finished control wire (Figure 5.17). Make two of them.

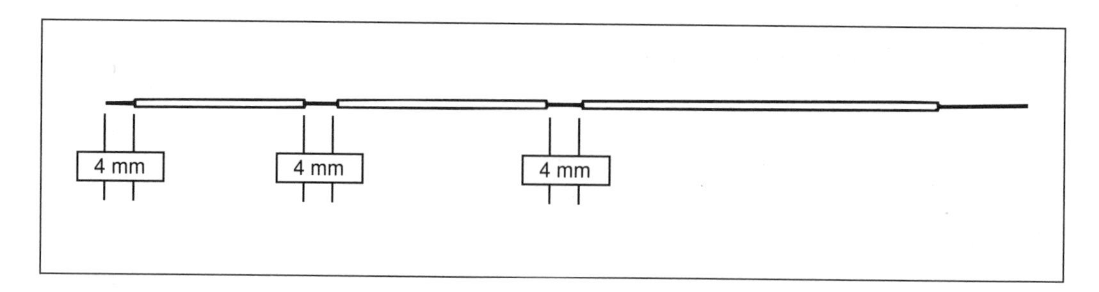

Figure 5.17. Finished control wire.

Do not install the control wires now. The control wires will be installed when the actuators are attached.

The Actuators

Stiquito is small and simple because it uses nitinol actuator wires (Figure 5.18).

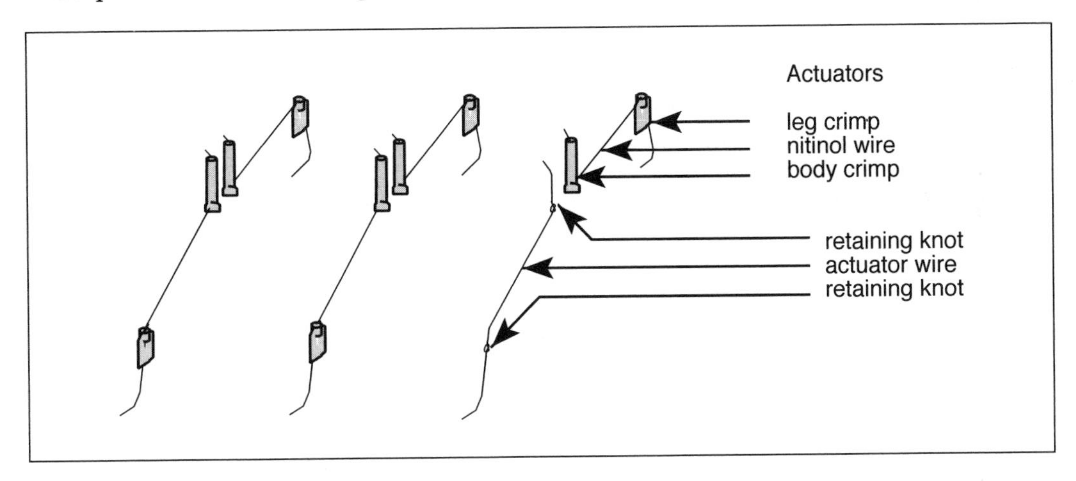

Figure 5.18. Actuators.

The nitinol wire translates the heat induced by an electric current into mechanical motion, replacing stepping motors, screws, and other components otherwise needed to make a leg move. The mechanical motion results from changes in the crystalline structure of nitinol (Figure 5.19). The crystalline structure is in a deformable state (the martensite) below the martensite transformation temperature, M_t. In this state the wire's length can change by as much as 10%. The nitinol wire is purchased as an expanded martensite (that is, a *trained* wire).

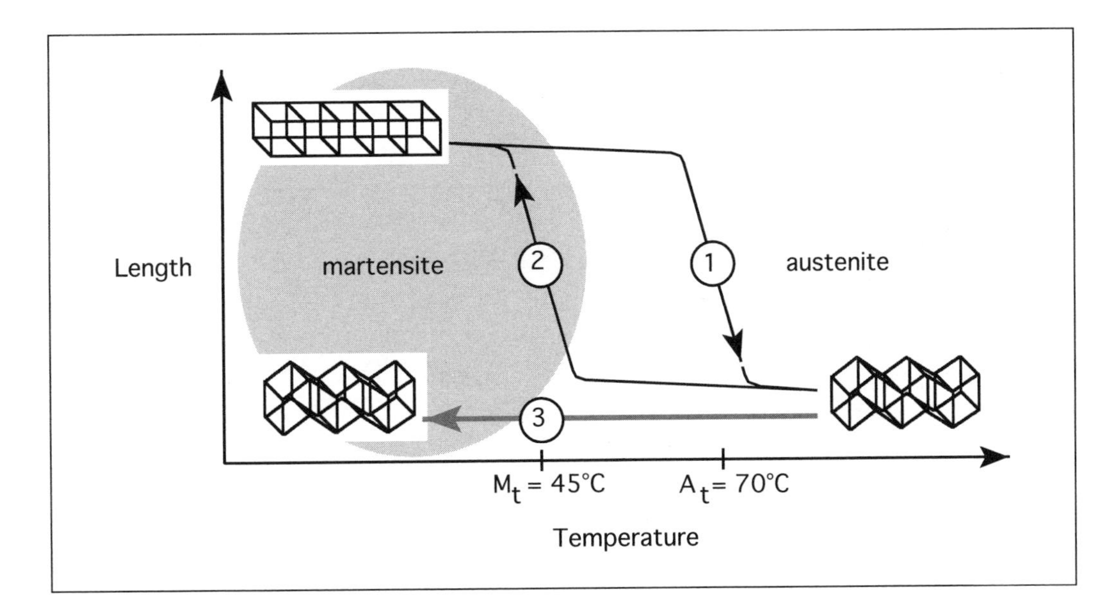

Figure 5.19. Changes in crystalline structure of nitinol.

When the wire is heated above the austenite transformation temperature A_t (1 in Figure 5.19), the crystalline structure changes to a strong and undeformable state (the austenite). As long as the temperature of the wire is kept slightly above A_t, the wire will remain contracted. During normal use of the nitinol wire, a recovery force, or tension, is applied while it is an austenite.

When the temperature falls below M_t the austenite transforms back into the deformable martensite (2 in Figure 5.19), and the recovery force pulls the wire back into its original, expanded form. If no recovery force is applied as the temperature falls below M_t, then the wire will remain short as it returns to the martensite (3 in Figure 5.19), although it can recover its original length by cycling again while a recover force is applied. If the wire is heated too far above A_t, then a new, shorter length results upon transformation to the martensite; the "memory" of the original, longer length cannot be restored.

Nitinol wire will operate for millions of cycles if it is not overheated and if a suitable recovery force is applied during each transformation. Stiquito's manual controller prevents overheating if used as directed. Autonomous controllers must limit the current supplied to the nitinol actuator wires to avoid overheating them. The music wire legs provide the correct recovery force.

The actuators, legs, and power bus combine to route power, provide the recovery force, and support the robot (Figure 5.20).

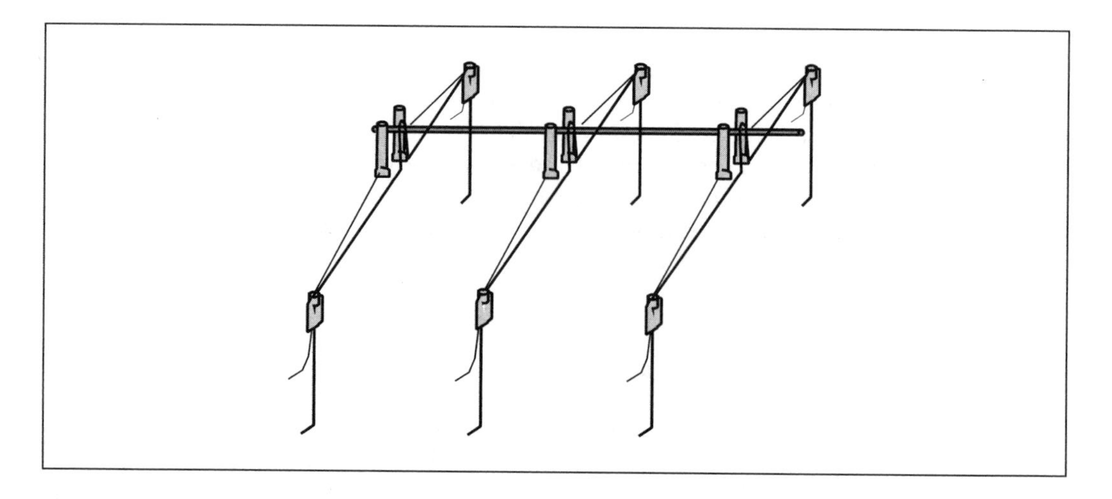

Figure 5.20. Actuators, legs, and power bus.

The following steps are needed to build the actuators, attach them to the legs and body, and form the ratchet feet.

Cut Nitinol Wires to Size—Begin making the actuators by cutting six 60 millimeter lengths (Figure 5.21) of nitinol wire (there will be extra nitinol wire left over). It might be preferable to use some of the extra nitinol in the kit, cutting six 70 millimeter lengths to make it easier to tie the retaining knots during installation of the leg crimps.

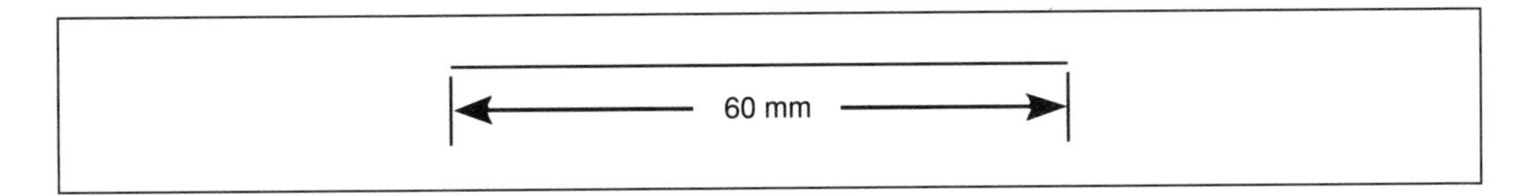

Figure 5.21. Nitinol wire.

Make Body and Leg Crimps—Next remove the 100 millimeter length of aluminum tubing from the kit. It will be used to make the leg and body crimps. Using the knife, cut six 9 millimeter body crimps and six 4 millimeter leg crimps from the aluminum tubing. Sand the ends of the crimps, then run the end of the knife through them to deburr the ends. (See Figure 5.22.)

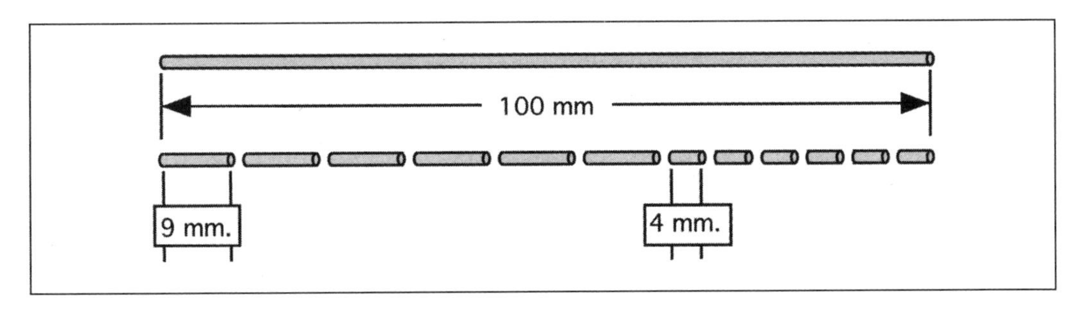

Figure 5.22. Body and leg crimps.

Attach Body Crimps to Nitinol Wires—Attach a body crimp to each length of nitinol wire. Tie a retaining knot in one end of the wire. Using the 600-grit sandpaper, lightly sand the nitinol wire at the knot to remove oxide and improve the electrical connection to the body crimp (Figure 5.23).

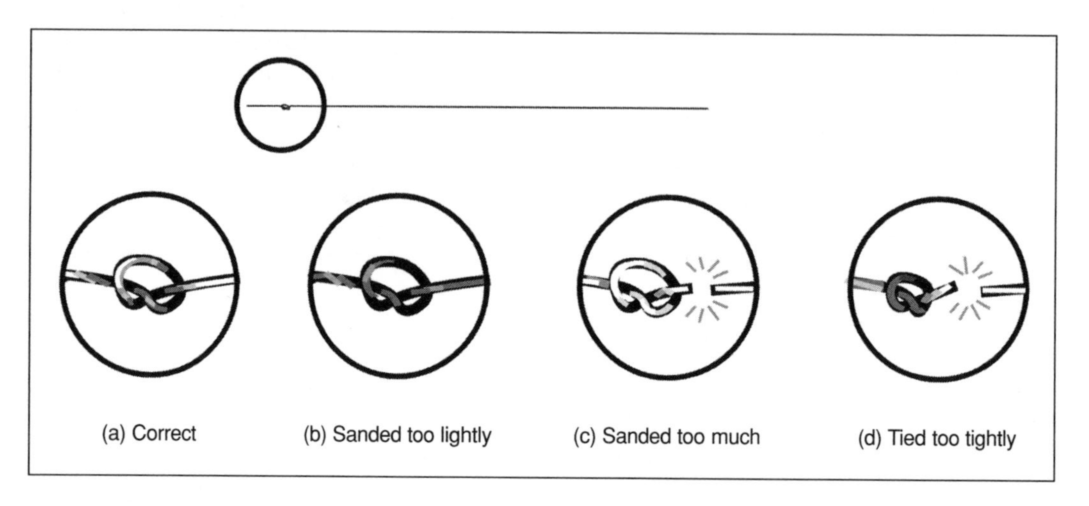

Figure 5.23. Tying and sanding retaining knot (all enlargements 10×).

Select a 9 millimeter body crimp, insert the knotted end of the wire into the body crimp, and pull it through the crimp; the knot must extend out the other end (Figure 5.24).

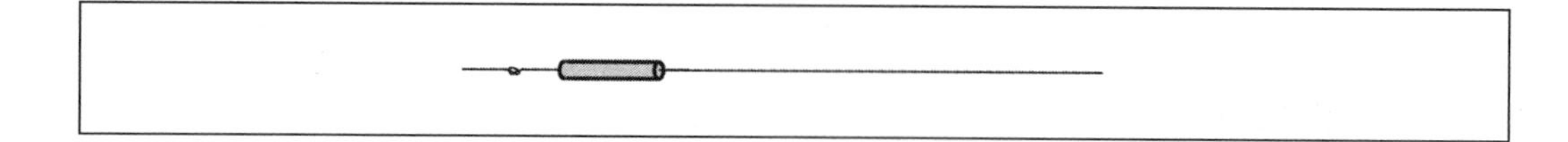

Figure 5.24. Knotted end of nitinol wire inserted into body crimp.

Turn the crimp until the nitinol wire is at either the right or left side of the crimp, then, using the needle-nose pliers, squeeze shut about 2 millimeters of the body crimp at the end opposite the knot. The unknotted wire should protrude from either the right or left side of the flattened part of the crimp, not the middle. Pull the knotted end back into the crimp until the retaining knot catches in the crimped end (Figure 5.25).

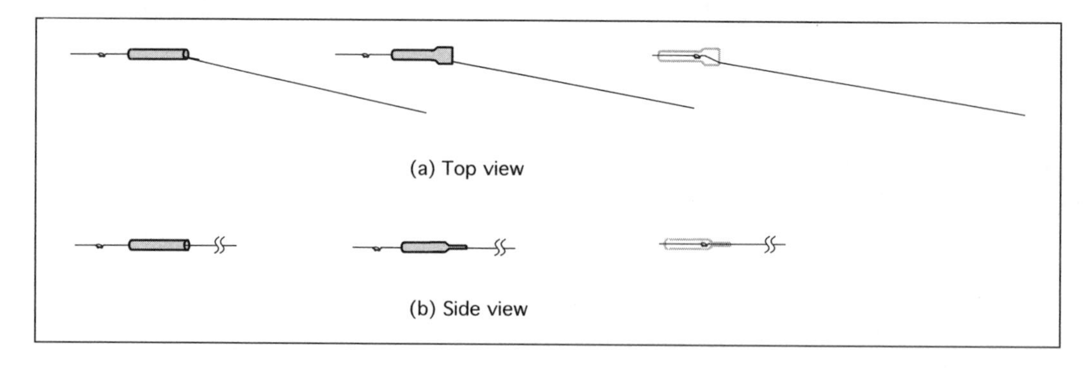

Figure 5.25. Attaching body crimp to nitinol wire.

Orient three body crimps so the nitinol wire exits from the left side of the flat, crimped end, and three so the nitinol wire exits from the right side (Figure 5.26). This ensures that the nitinol wire will protrude as far to the rear as possible from the music wire leg when the actuator is attached to the body.

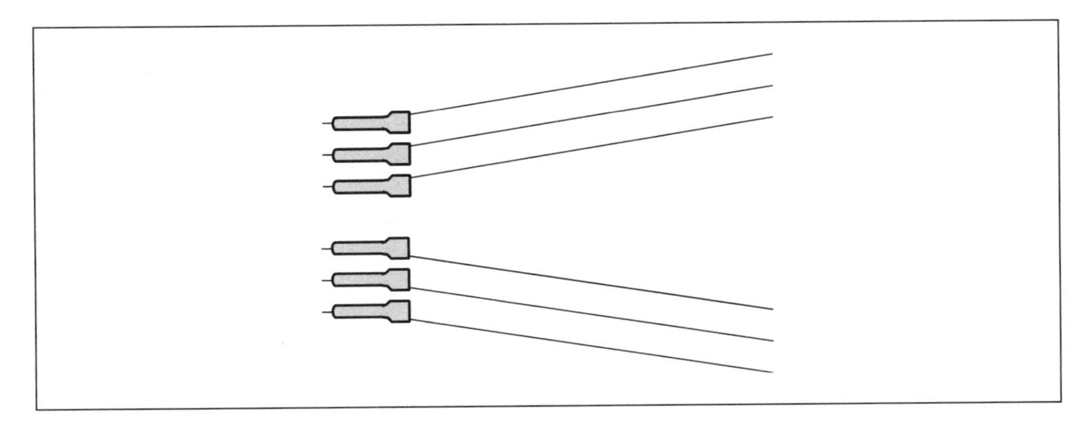

Figure 5.26. Left- and right-oriented body crimps.

Next, retaining the left- and right-handed orientation of each body crimp, grasp the flat end of each body crimp with the needle-nose pliers and bend it upward to a 45-degree angle (Figure 5.27).

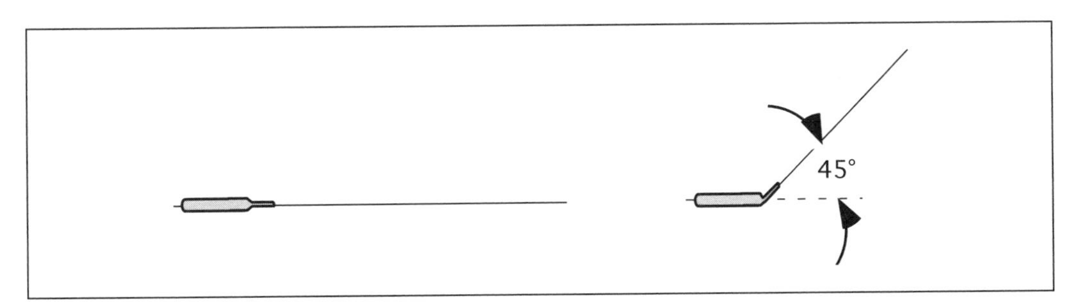

Figure 5.27. Bending the body crimp.

There should now be three left and three right body crimps with nitinol wire attached (Figure 5.28).

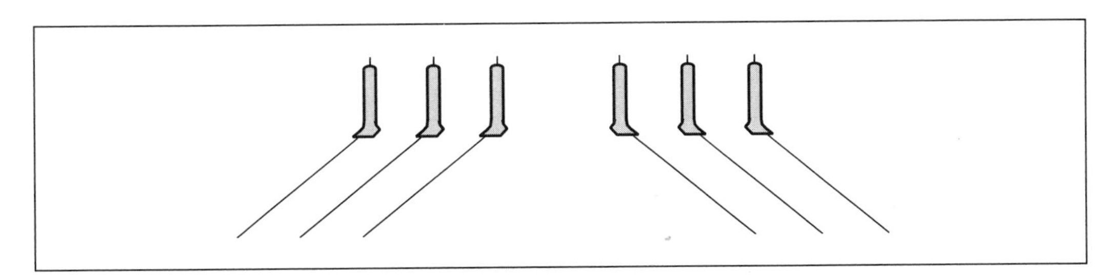

Figure 5.28. Finished left- and right-handed body crimps.

Insert Control Wires and Body Crimps into Body Crimp Holes—Viewing the body with the bottom side facing up and the front of the body at the tip, use left-handed body crimps on the left side of the body and right-handed body crimps on the right side. This will put the attachment point for each actuator wire as far to the rear of the leg as possible (Figure 5.29).

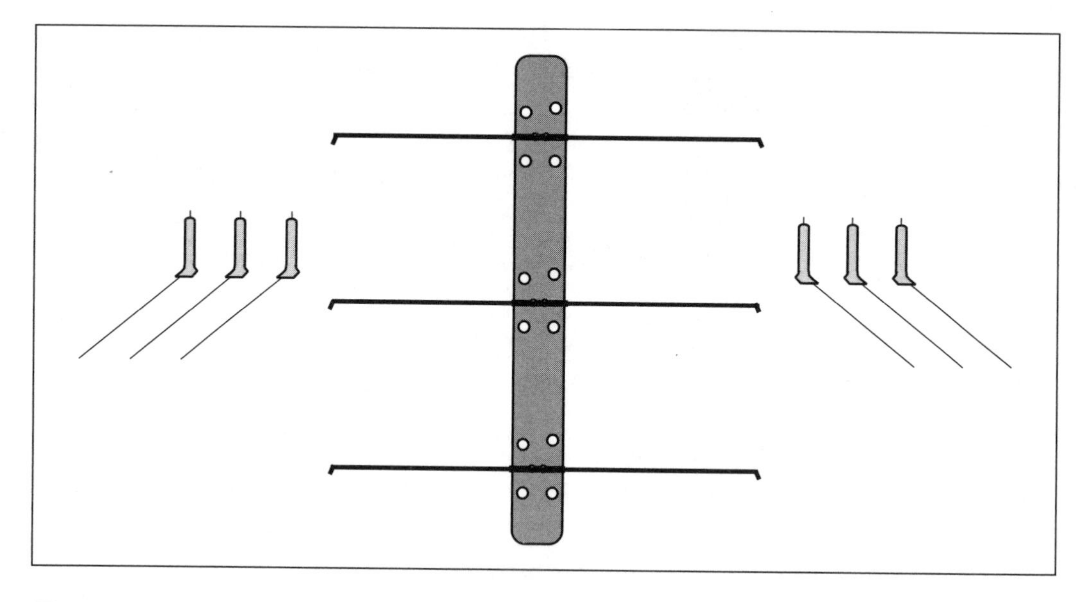

Figure 5.29. Orienting body crimps to bottom side of the body (front of body at top).

Beginning at the front of the body, insert the end of one control wire so that the bare wire enters the front-left body crimp hole at the V-groove (Figure 5.30).

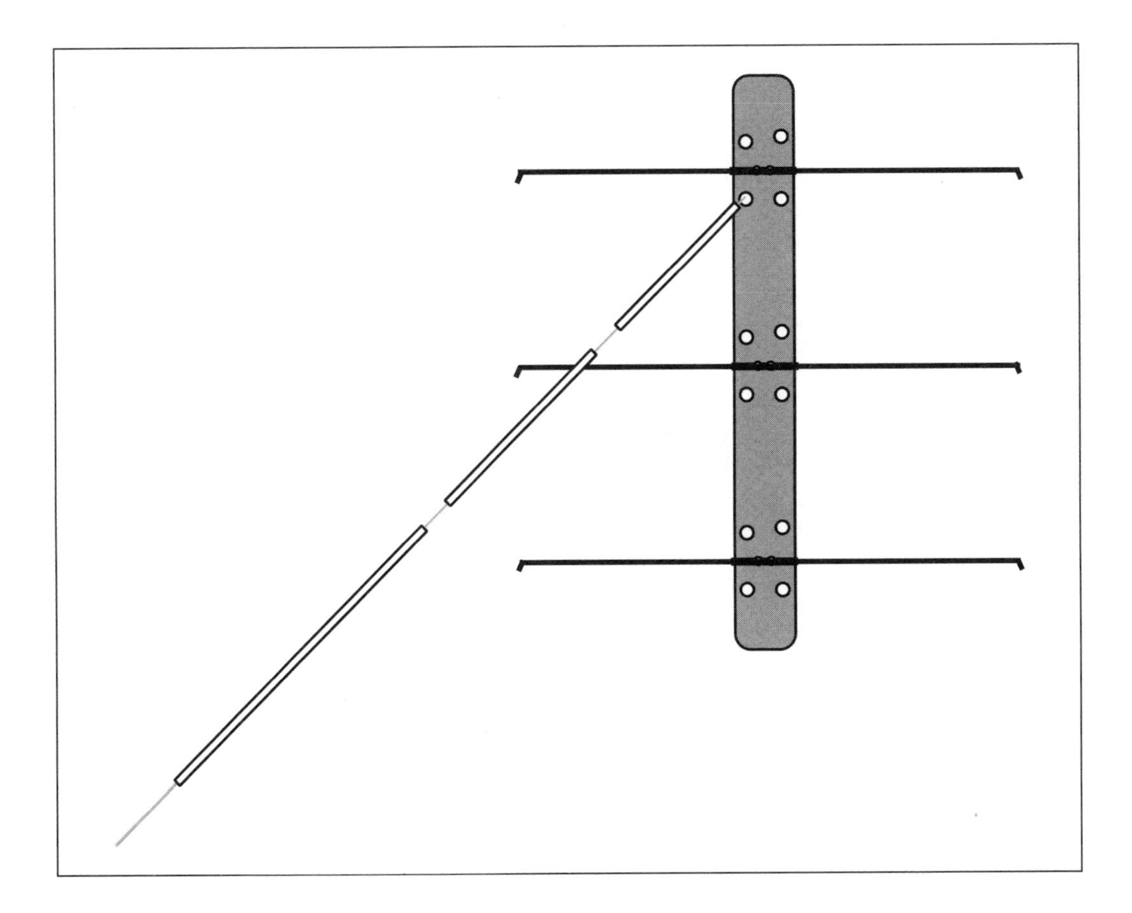

Figure 5.30. Inserting control wire.

Orient and insert a left-handed body crimp into the body crimp hole, securing the control wire (Figure 5.31).

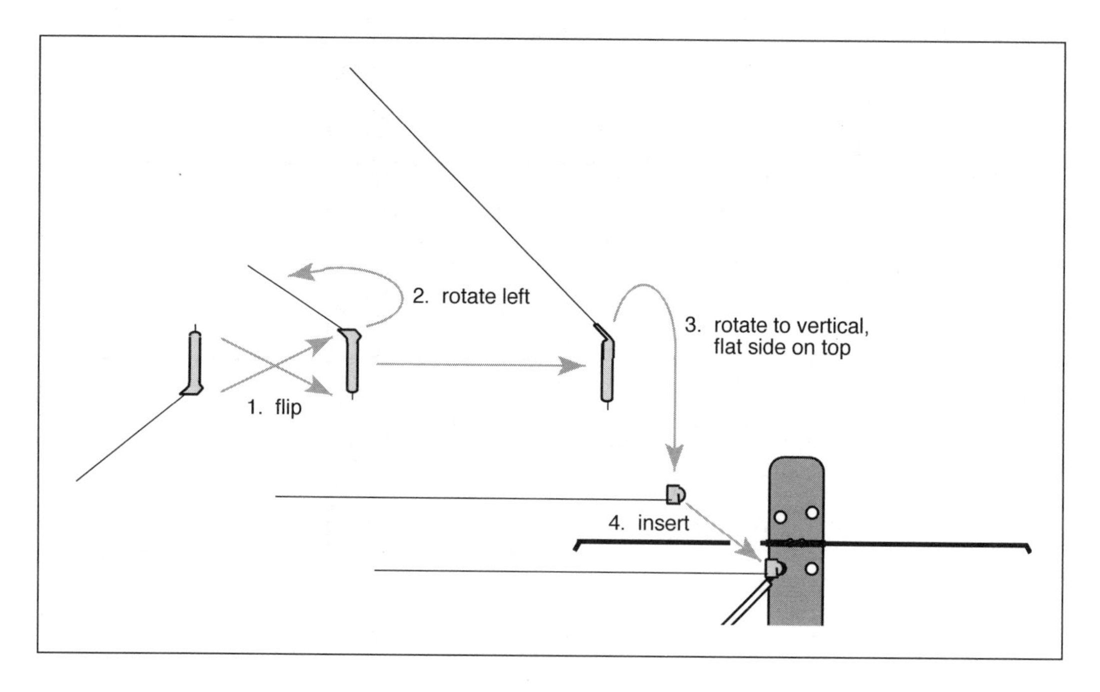

Figure 5.31. Attaching body crimp.

Continue to string the control wire and secure it with body crimps, hardwiring a tripod gait. The control wire should be run through the front-left, middle-right, and rear-left body crimp holes (Figures 5.32 and 5.33).

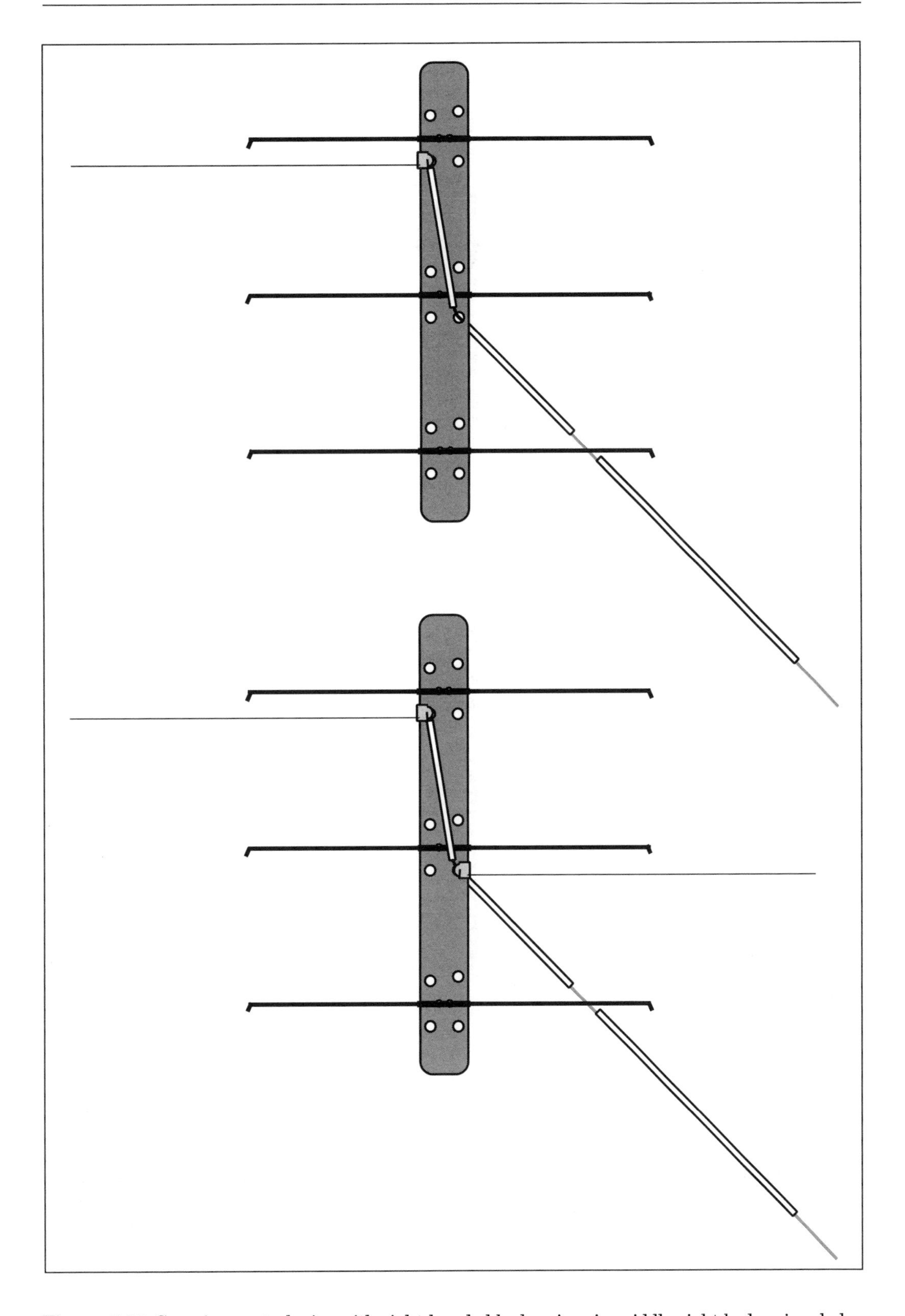

Figure 5.32. Securing control wire with right-handed body crimp in middle-right body crimp hole.

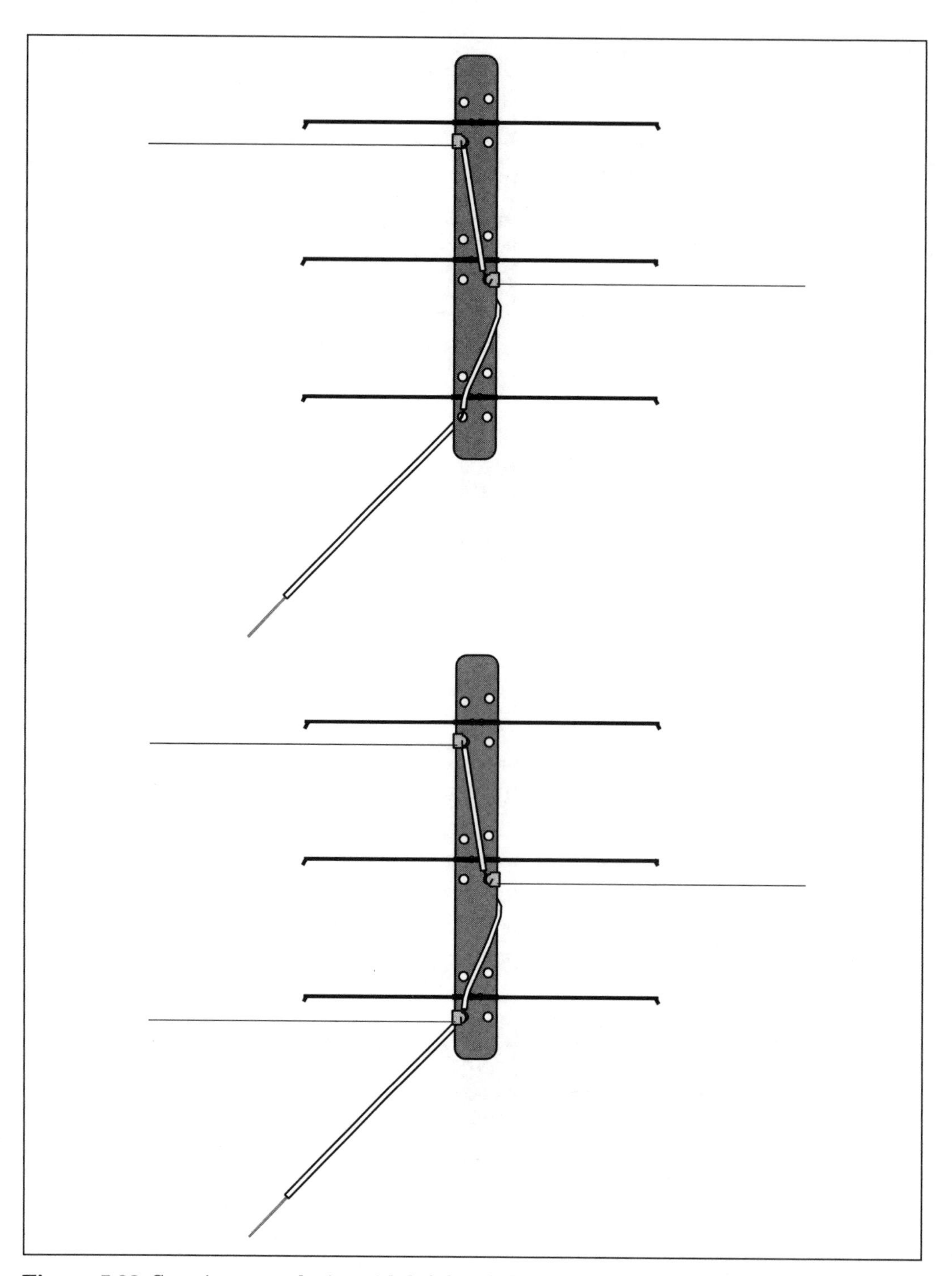

Figure 5.33. Securing control wire with left-handed body crimp in rear-left body crimp hole.

When installing the rear crimp, wrap the control wire around the body so that it enters the body crimp hole from the bottom side of the body.

Following a similar sequence, install the second control wire in the opposite holes. See Figure 5.34.

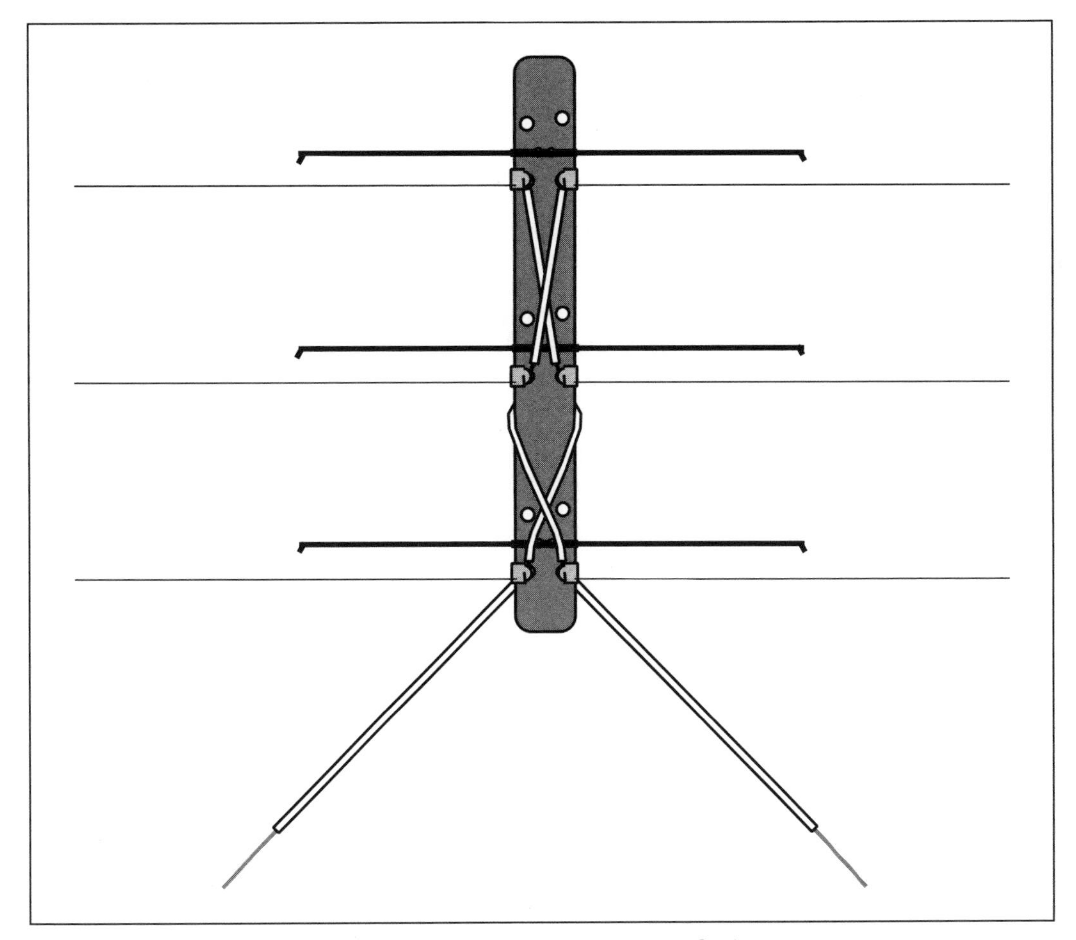

Figure 5.34. Body with attached body crimps securing control wires.

Finish the control wire and body crimp assembly by bending the 45-degree tabs on the body crimps flat against the body using the needle-nose pliers (Figure 5.35). This causes the finished nitinol actuator wires to pull the leg backward and slightly downward during operation.

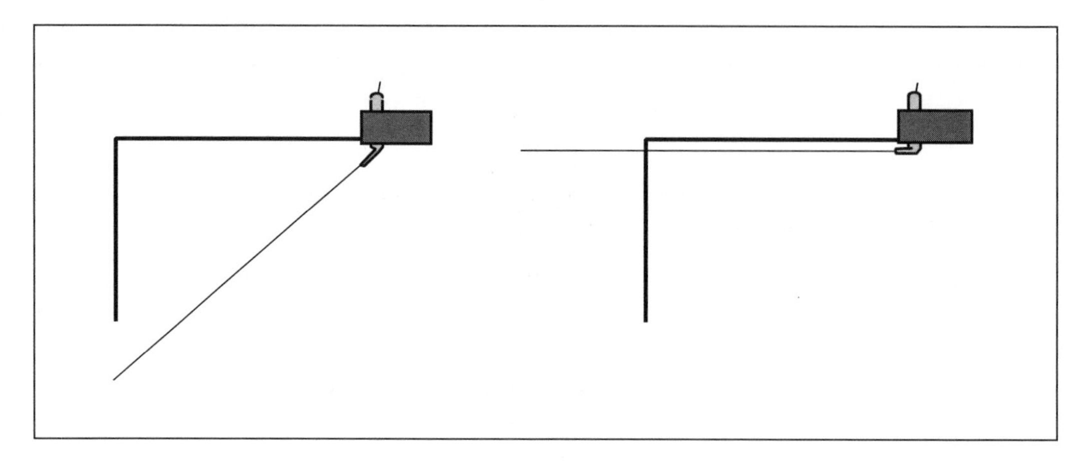

Figure 5.35. Finishing the body crimps (end view).

Attach Leg Crimps—For each nitinol wire actuator, position a knot at the knee near the vertical joint by tying an overhand knot around the vertical joint (Figure 5.36). Tighten the knot to a 0.1 mm loop with the pliers.

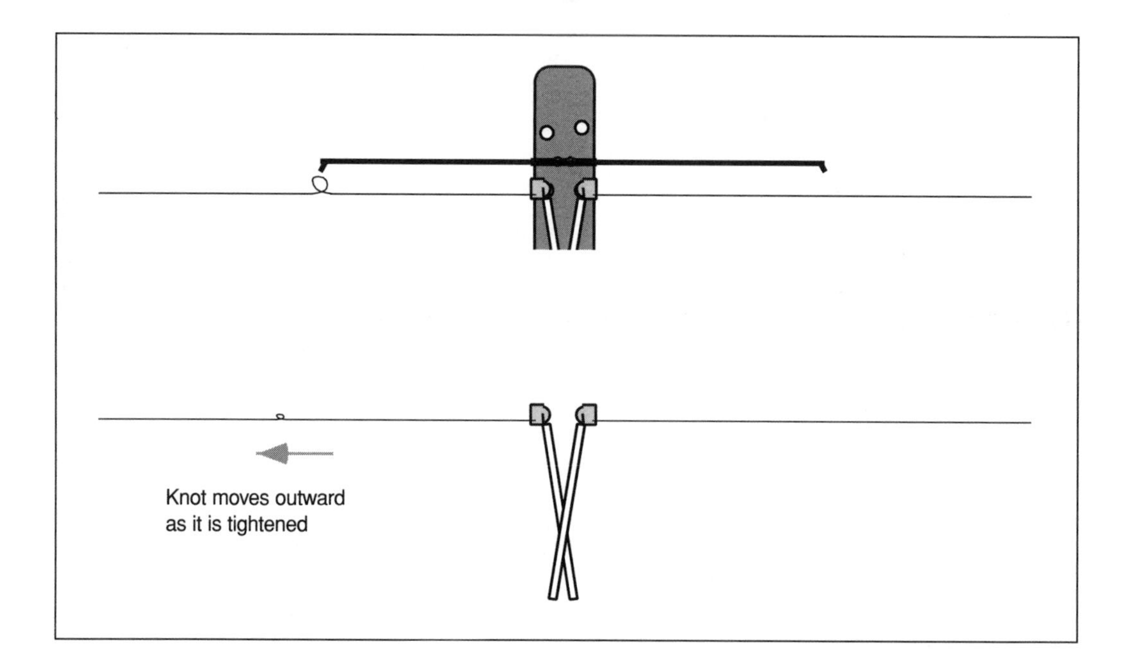

Figure 5.36. Locating leg crimp knot.

Sand the knot and the vertical joint near the knee with 600-grit sandpaper. Select a 4 millimeter leg crimp and slip it onto the vertical joint and slide it up to the knee (Figure 5.37).

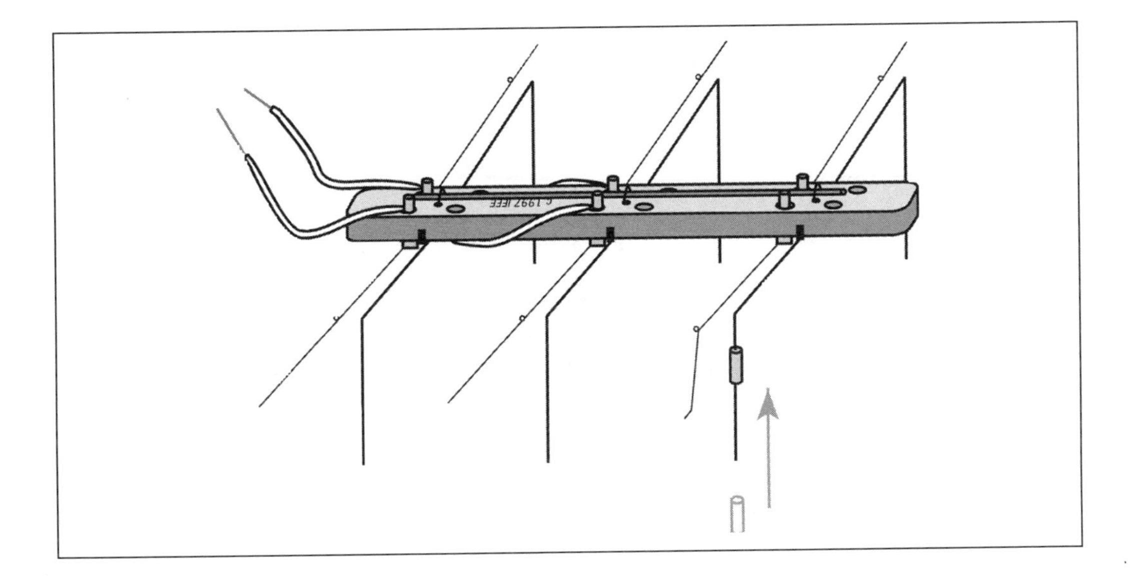

Figure 5.37. Slip crimp onto leg.

Thread the nitinol wire through the crimp (Figure 5.38).

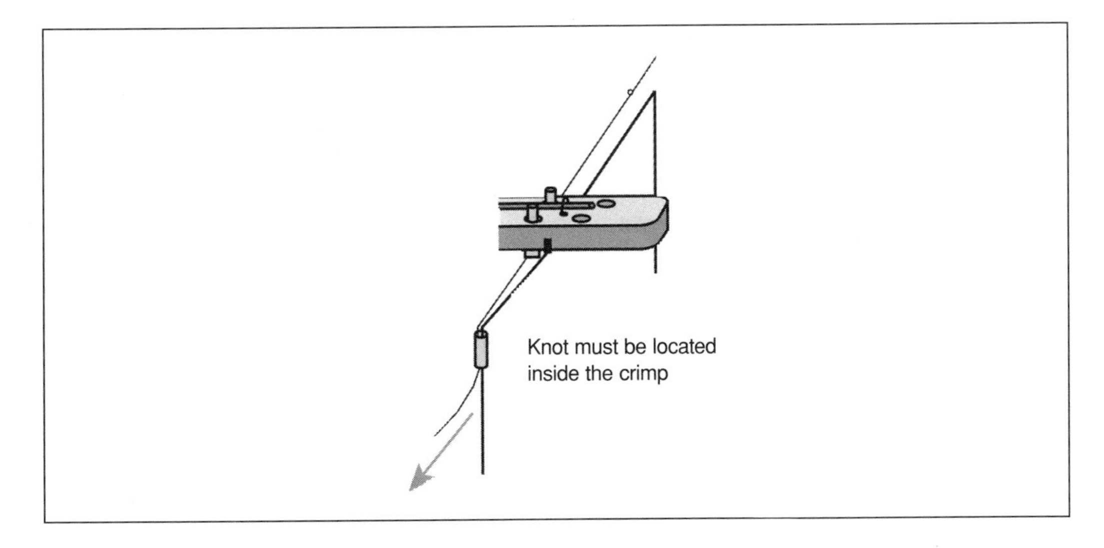

Figure 5.38. Threading wire through crimp.

Pull the wire taut so there is a slight backward bend in the horizontal joint. The slight bend ensures there is no slack in the nitinol actuator wire. There is not enough tension if there is no bend in the leg. There is too much tension if the leg is bent more than 2 millimeter backward at the knee. See Figure 5.39.

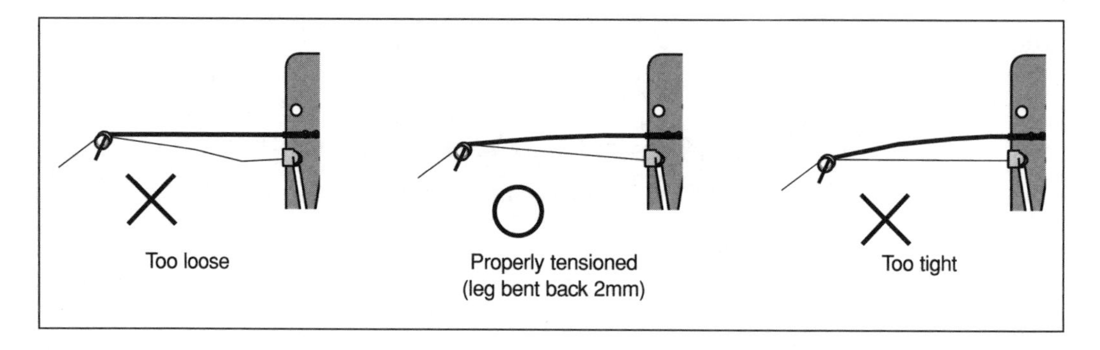

Figure 5.39. Correct tension (O) and incorrect tension (X).

The nitinol actuator wire should enter the crimp at the point nearest the body. The knot should be inside the crimp. When these conditions are met, use the needle-nose pliers to make a loose crimp by squeezing the bottom 3.5 millimeter of the crimp just enough to catch the knot (Figure 5.40).

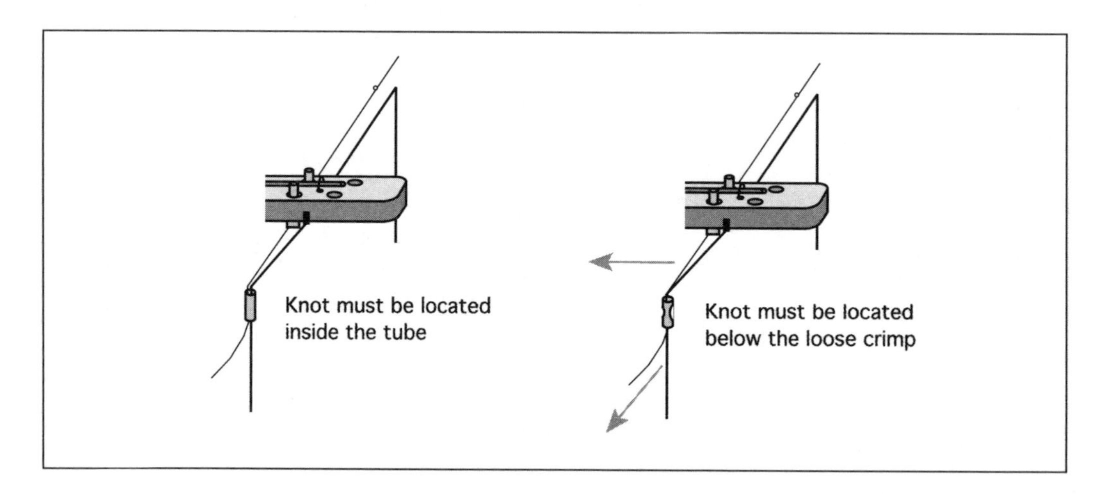

Figure 5.40. Making the loose crimp.

Crimping changes the tension in the actuator wire and might allow it to become slack. Adjust the tension while the crimp is loose by pulling on the wire (the horizontal joint should be bent slightly backward). When the actuator wire is taut, hold it in place and squeeze the crimp shut tightly (Figure 5.41).

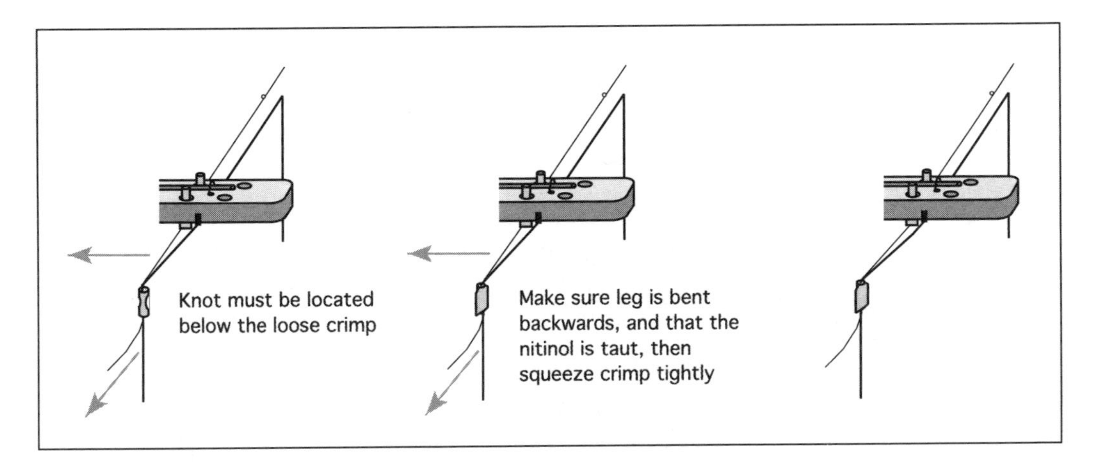

Figure 5.41. Making the tight crimp.

Measure the resistance from the leg near the clip groove to the body crimp where it protrudes above the top of the body. The initial resistance should be between 5 and 7 ohms. The resistance will increase to between 15 and 25 ohms as the connections age. Test the operation of the actuators by applying current from two 1.5-volt AA cells at the leg near the body and the body crimp for no more than 0.5 seconds to prevent the actuator wire from overheating. The leg should immediately bend backward 3 to 7 millimeters as measured at the vertical joint, and then return to its original position. See Figure 5.42.

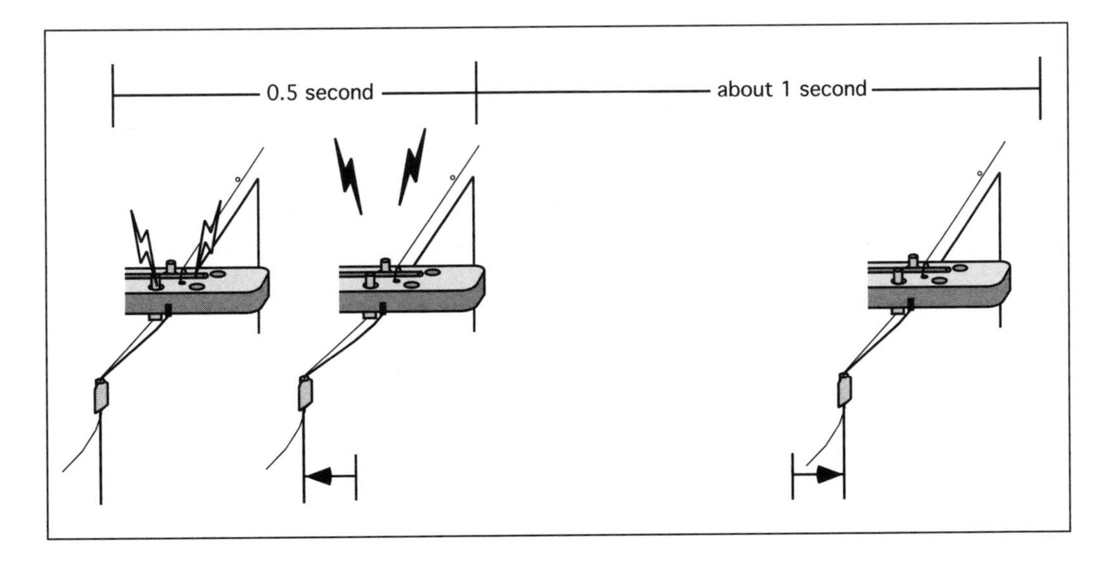

Figure 5.42. Testing the actuator.

Test the leg and actuator assembly after installing each leg crimp.

If the test is successful, continue to attach and test the remaining leg crimps. See the views of finished actuators in Figures 5.43 and 5.44.

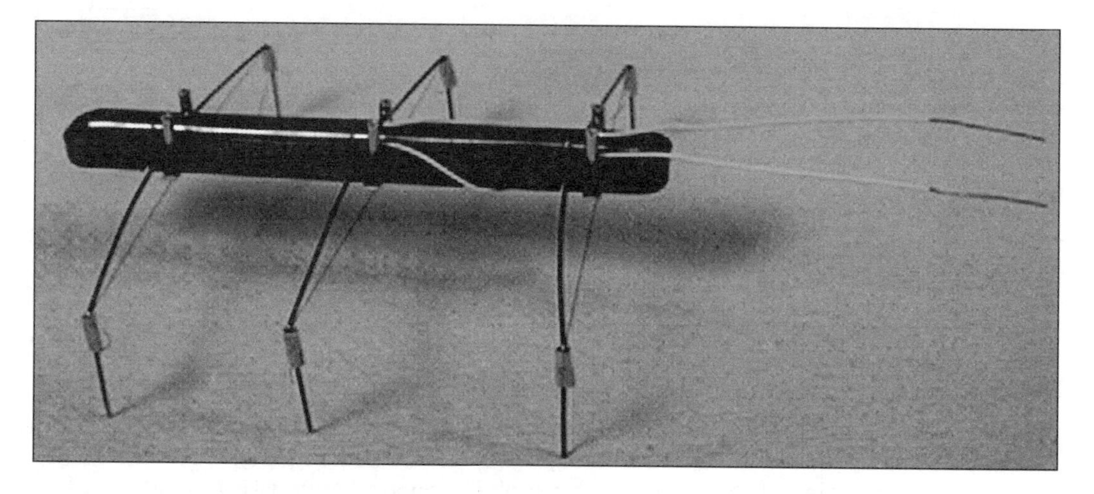

Figure 5.43. Finished actuators (side view).

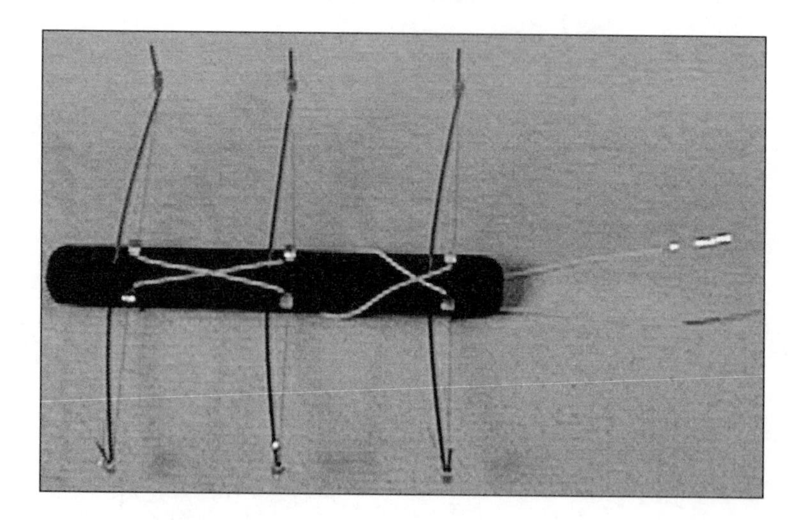

Figure 5.44. Finished actuators (bottom view).

The Ratchet Feet—Be sure to test the actuators before making the ratchet feet in case the leg crimps must be replaced. Form an ankle and ratchet foot by bending the tip of each vertical joint backward and slightly outward. The ratchet foot should be about 2 millimeters long and make a 110-degree angle downward from the vertical joint. The ankle should face toward the front of the robot. See Figure 5.45.

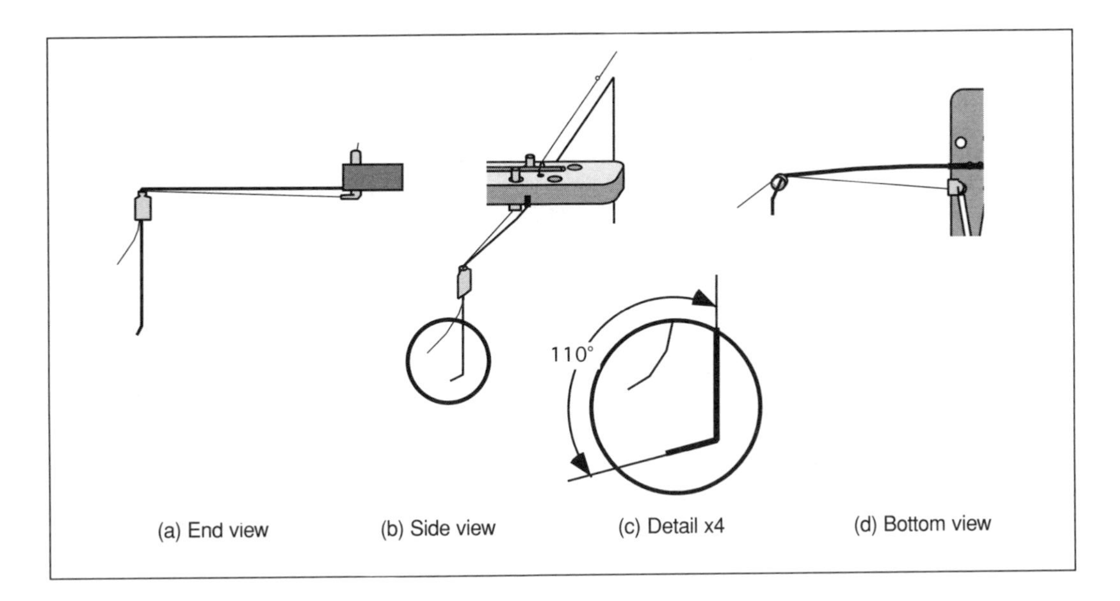

Figure 5.45. Ratchet foot.

This completes the robot (Figures 5.46 and 5.47).

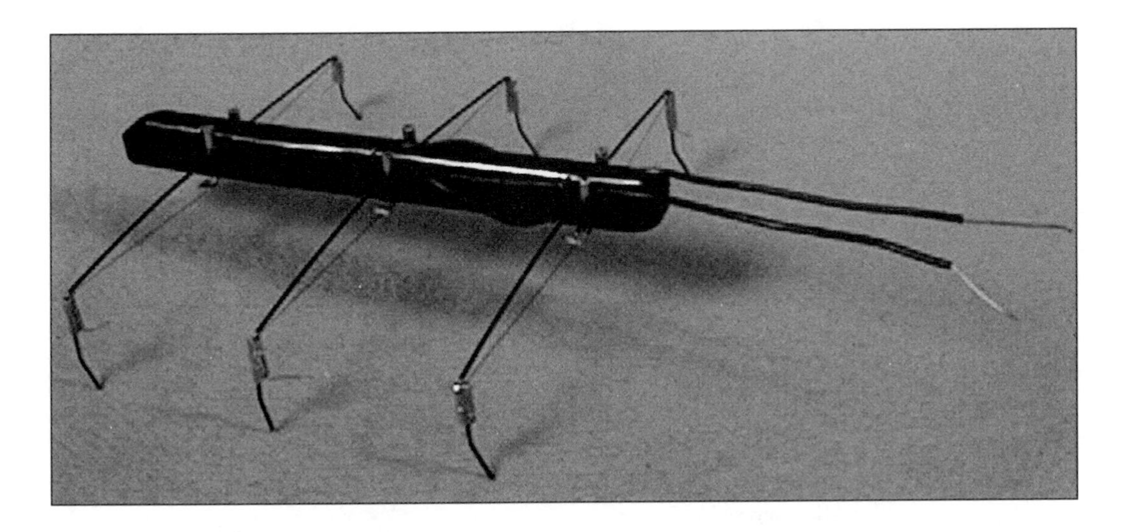

Figure 5.46. Side view of finished robot.

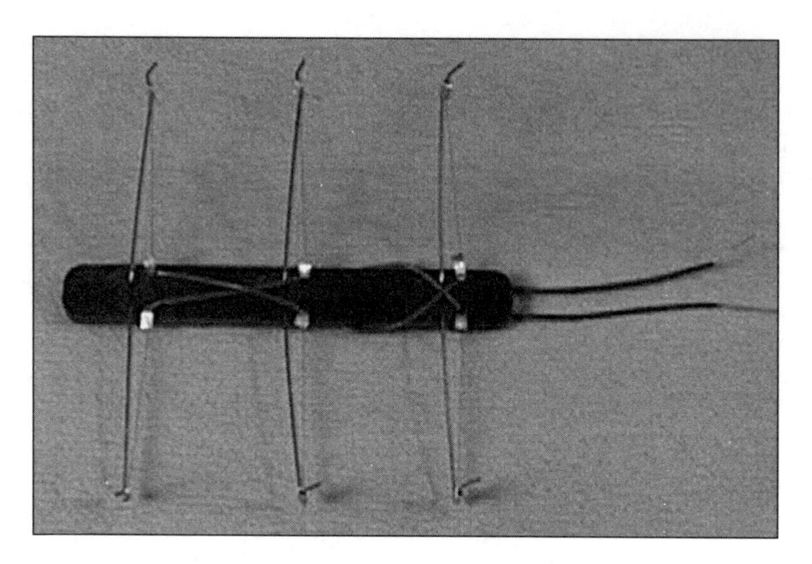

Figure 5.47. Finished actuators (bottom view).

TAKING STIQUITO FOR A WALK

Stiquito is hardwired to walk in a tripod gait. Other gaits can be hardwired by removing the body crimps and rewiring the control wires. Each leg can be wired separately for more complex gaits, especially when using a hardware controller on an autonomous robot.[9]

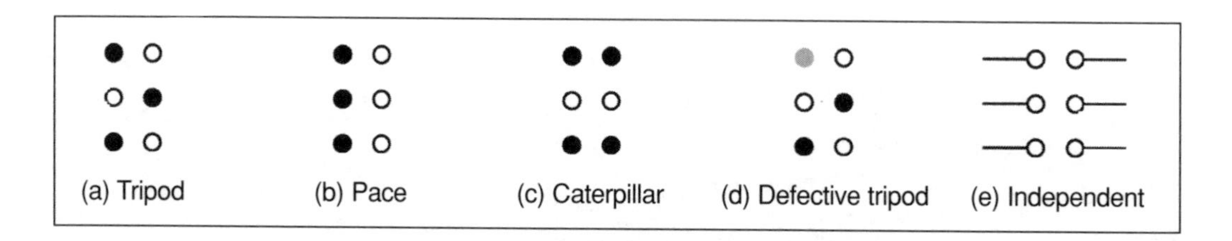

Figure 5.48. Gaits.

This book provides instructions for three types of controllers:

- Chapter 6—Manual Controller. The kit provided in the book contains the materials needed to make the manual controller. This controller is simply two on and off switches: they turn on and off current to the nitinol wires, causing the legs to move. This controller provides a quick way to test your Stiquito robot and ensure Stiquito will walk correctly.
- Chapter 7—PC-Based Controller. You can control the walking gait of your Stiquito robot by writing a program on your personal computer. The program uses the

parallel port of the PC, an attached circuit board, and an additional power source to turn on and off current to the nitinol wires. The test board is built to verify your programs with LEDs before you attach your robot. A kit containing all of the supplies you need to build this controller can be purchased from one of the suppliers listed in Appendix B.

- Chapter 8—Analog Controller. You can build a circuit board that mounts on top of the Stiquito robot and directs the robot to walk in a tripod gait. This circuit will feed current to the nitinol on a periodic basis, which you can adjust. This setup allows Stiquito to be an autonomous robot. Again, a kit containing all of the supplies you need to build this controller can be purchased from one of the suppliers listed in Appendix B.

Please refer to these chapters for specific descriptions and building instructions. Please remember that the circuit materials described in Chapters 7 and 8 are not included in this book.

EXERCISES

1. Define the following terms:
   a. actuator
   b. bus
   c. controller
2. If Stiquito walks at 10 centimeters per minute, what is its speed in kilometers per hour? How does this compare with an automobile, which can travel at 100 kilometers per hour?
3. The length of nitinol can change as much as 10% when heated. Write the formula to determine the percentage change in length, using units of its cooled and heated lengths.
4. This exercise will determine the amount of contraction of your nitinol wire. Do this by measuring the nitinol at rest and when contracted. Because you should not allow the wire to contract too long, measure the contracted length by quickly marking paper with a pencil:
   a. Complete building Stiquito up until you attached one nitinol wire to one leg and tested your work (after the test of Figure 5.42).
   b. Using a ruler graduated in millimeters, measure the length of the nitinol wire when it is cool.
   c. Next, hold the robot firmly down on a piece of paper on a table, legs up.
   d. Apply 3 volts to the leg and mark on the paper the point where the nitinol stopped contracting (mark the paper at the knee).
   e. Remove the voltage.
   f. Without moving the robot, mark the paper under where the nitinol leaves the body crimp.
   g. Move the robot, measure the distance (which is the contracted length).

h. Calculate the percentage of contraction.

5. Describe changes you would need to make to Stiquito for independent leg control.
6. Describe changes you would need to make to Stiquito to allow it to walk with two degrees of freedom.
7. Observe your Stiquito walking. Describe how it mimics the movement of living things. What are the similarities? What are the differences? Are there any classes of living things that closely resemble Stiquito?
8. How can Stiquito be changed to walk better?

REFERENCES

1. Dynalloy, Inc. *Technical characteristics of Flexinol™ actuator wires.*
2. Gilbertson, R. G. 1992. *Working with shape memory wires.* San Leandro, Calif.: Mondo-Tronics.
3. Mills, J. W. 1992. Area-efficient implication circuits for very dense Lukasiewicz logic arrays. *Proc. 22nd Int'l Symp. Multiple-Valued Logic.* Sendai, Japan: IEEE Press.
4. Mills, J., and C. Daffinger. 1990. An analog VLSI array processor for classical and connectionist AI. *Application Specific Array Processors.* Princeton, N.J.: IEEE Press.
5. Brooks, R. 1990. A robot that walks: Emergent behaviors from a carefully evolved network. *Neural Computation 1,* no. 2:253–262.
6. Beers, R. 1991. An artificial insect. *American Scientist 79* (Sept.–Oct.): 444–452.
7. Wilson, E. O. 1975. *Sociobiology: The new synthesis.* Harvard University Press.
8. Ballard, D. H., and C. M. Brown. 1982. *Computer Vision.* Englewood Cliffs, N.J.: Prentice Hall.
9. See Chapters 7 through 9 for examples.

Chapter 6

A Manual Controller for the Stiquito Robot

Jonathan W. Mills

INTRODUCTION

A simple manual controller to test Stiquito's walk is described in this chapter. The kit provided in the book contains the materials needed to make the manual controller. This controller is simply two on/off switches: they turn on and off current to the nitinol wires, causing the legs to move. This controller provides a quick way to test your Stiquito robot and ensure Stiquito will walk correctly.

The final section of this chapter contains troubleshooting steps. This section lists the symptoms and causes of some common problems that might be encountered, based on experience with this robot.

Before you start building, gather the remaining Stiquito kit materials, your tools, and the 9-volt battery you purchased.

BUILDING THE MANUAL CONTROLLER

A manual controller is a good way to test hardwired gaits. It is simple and almost foolproof. If used as directed, it will prevent overheating that could damage the nitinol actuator wires. The manual controller is a pair of normally open switches with a terminal to attach a 9-volt cell. Pressing the switches alternately will power the legs, causing Stiquito to walk. If Stiquito walks with the manual controller, then the control circuits described in later chapters should work equally well.

Make the manual controller by starting with the provided plastic handpiece (Figure 6.1).

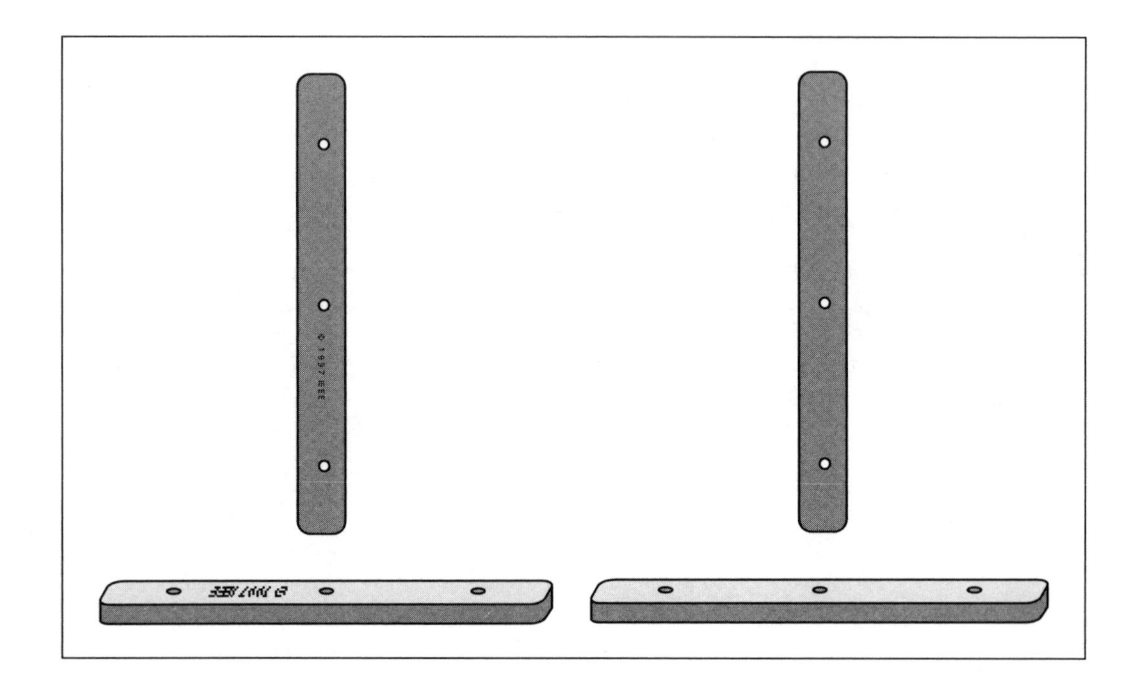

Figure 6.1. The molded plastic manual controller handpiece.

Cut two 120-millimeter lengths of 28 AWG copper wire wrap wire from the kit. Strip 12-millimeters of insulation from each end (Figure 6.2).

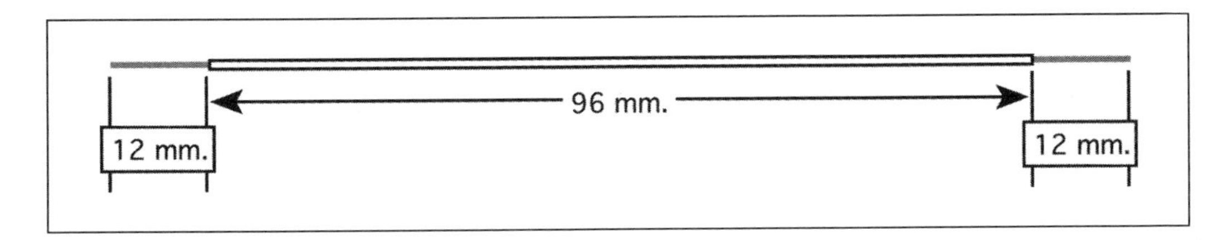

Figure 6.2. Control wire.

Bend one end of each wire in half, leaving a sharp V-bend of bare wire 6 millimeters long (Figure 6.3).

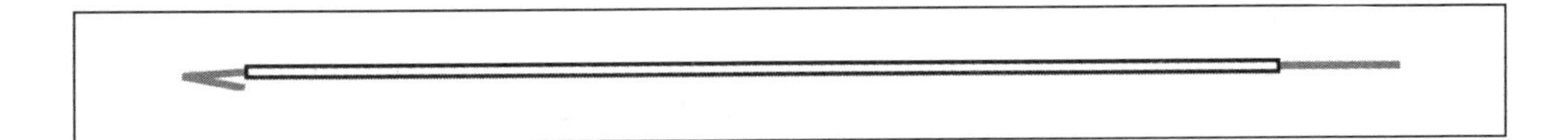

Figure 6.3. V-bend.

Remove the second 100-millimeter piece of aluminum tubing from the kit. Cut three 9-millimeter crimps. Two will be contact crimps, and one will be a power crimp. Cut six 4-millimeter connecting crimps. Approximately 40 millimeters of kit tubing remains for spare crimps. See Figure 6.4.

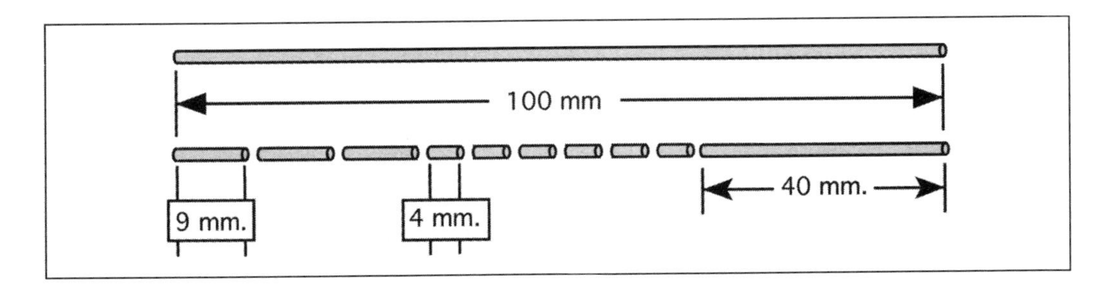

Figure 6.4. Contact, power, and connecting crimps (spare tubing shown).

Make two contact crimp assemblies. Into each of two 9-millimeter crimps, insert a V-bend until the tip protrudes slightly beyond the crimp. Using needle-nose pliers, crimp 2 millimeters of the end with the slightly protruding V-bend. See Figure 6.5.

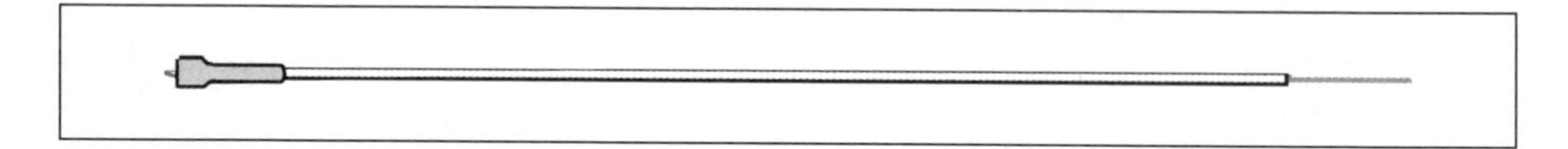

Figure 6.5. Contact crimp assemblies.

Bend the flat part of the crimp to a 45-degree angle (Figure 6.6).

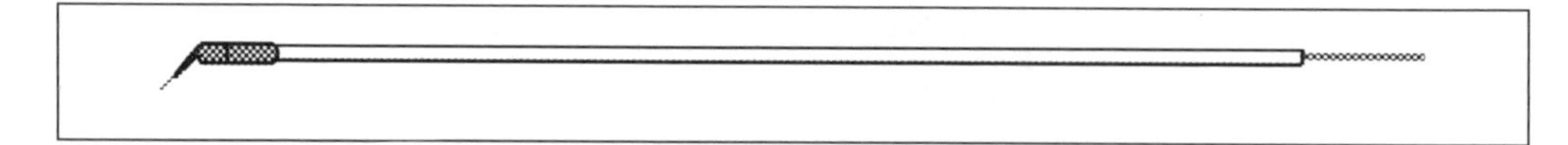

Figure 6.6. 45-degree bend in contact crimp.

Next make one power crimp assembly. Select the 9-volt terminal. If necessary, remove 5 millimeters of insulation from the end of each wire attached to the terminal. Insert one of the wires into the remaining 9-millimeter crimp all the way to the insulation and crimp 2 millimeters at the end nearest the insulation. See Figure 6.7.

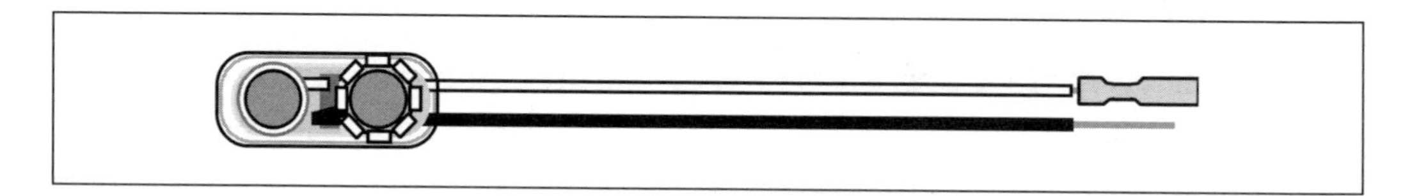

Figure 6.7. Power crimp assembly.

Take one 85-millimeter length of 0.020 inch music wire from the kit. Bend the wire in the center to a 15-degree angle to make a V-clamp (Figure 6.8).

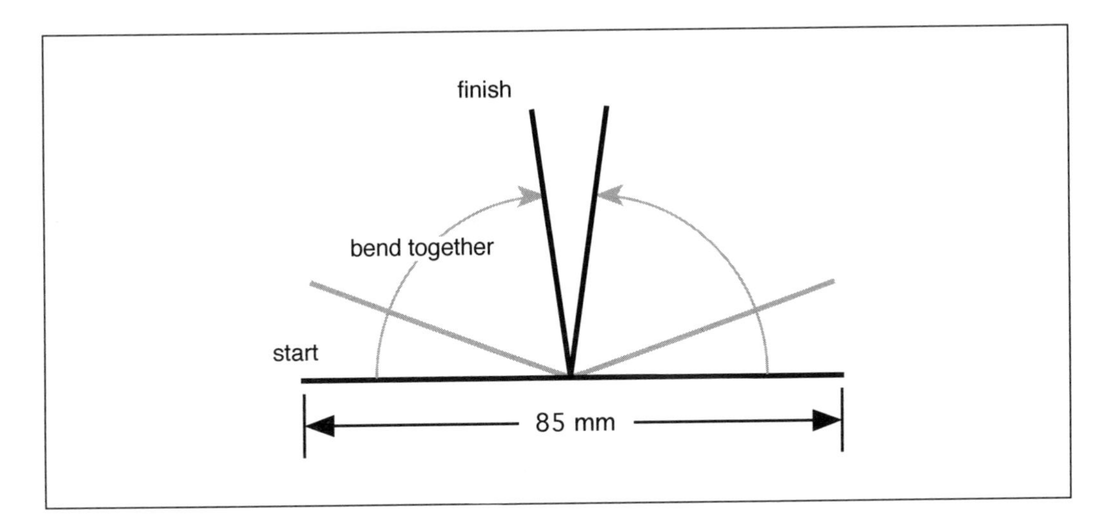

Figure 6.8. V-clamp.

Insert the music wire into the center hole in the plastic handpiece. Make sure there is some space between the wires near the bend (Figure 6.9).

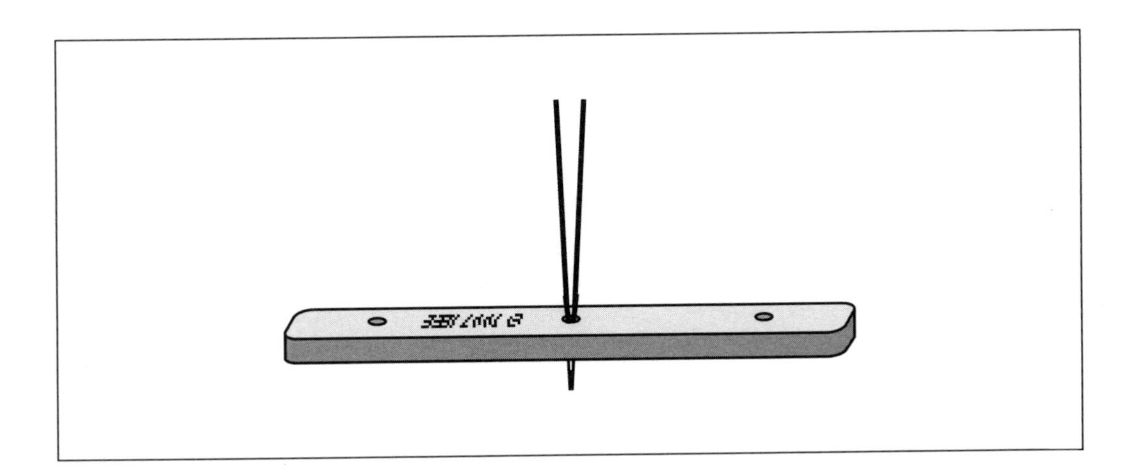

Figure 6.9. Insert music wire.

Insert the power crimp between the wires at the bend of the music wire (Figure 6.10). Center the crimp at the bend.

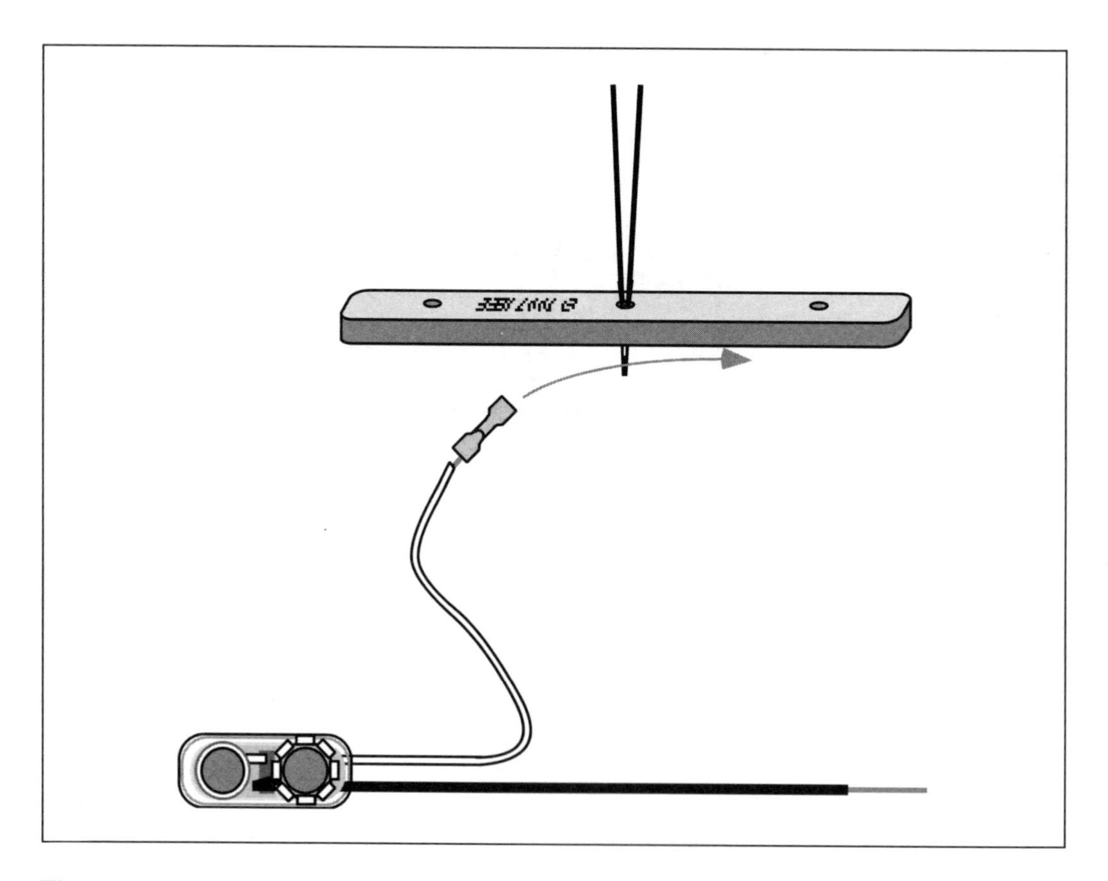

Figure 6.10. Insert power crimp.

Place the manual controller with the power crimp down on a hard surface. Firmly press down on the plastic handpiece with your thumbs, as close to the music wire as possible. Next, bend the music wire to the outside of the handpiece (Figure 6.11 and 6.12).

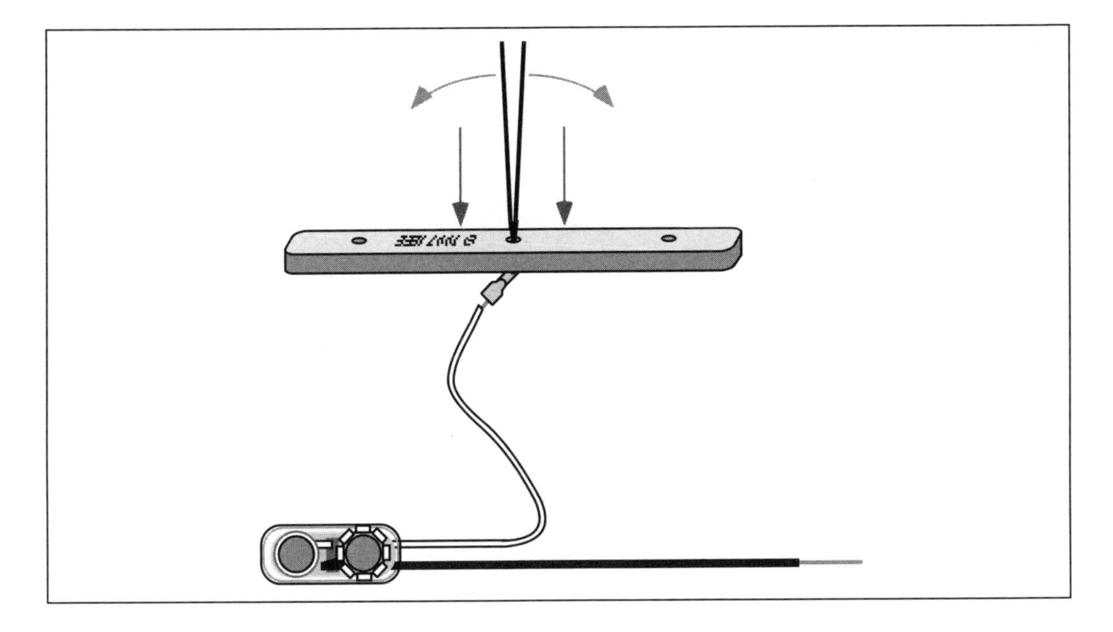

Figure 6.11. Bending the V-clamp outwards. Press down firmly to catch crimp tightly in "V" while bending music wire outward.

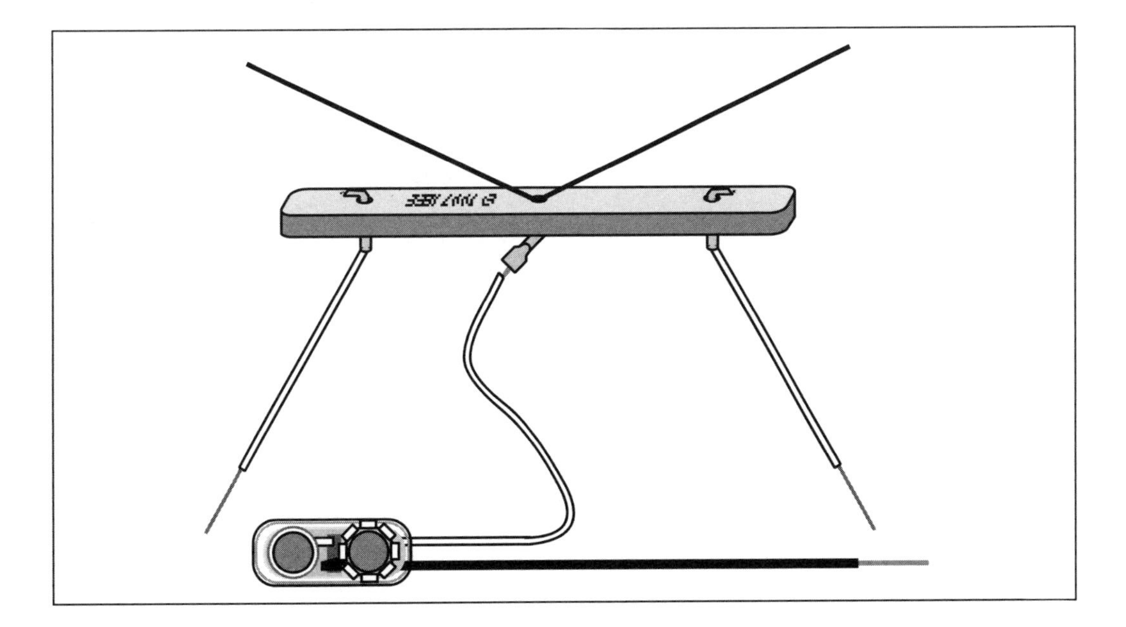

Figure 6.12. Insert and bend contact crimp assemblies.

Bend each end of the V-clamp back on itself to prevent poking yourself in the finger when using the controller (Figure 6.13).

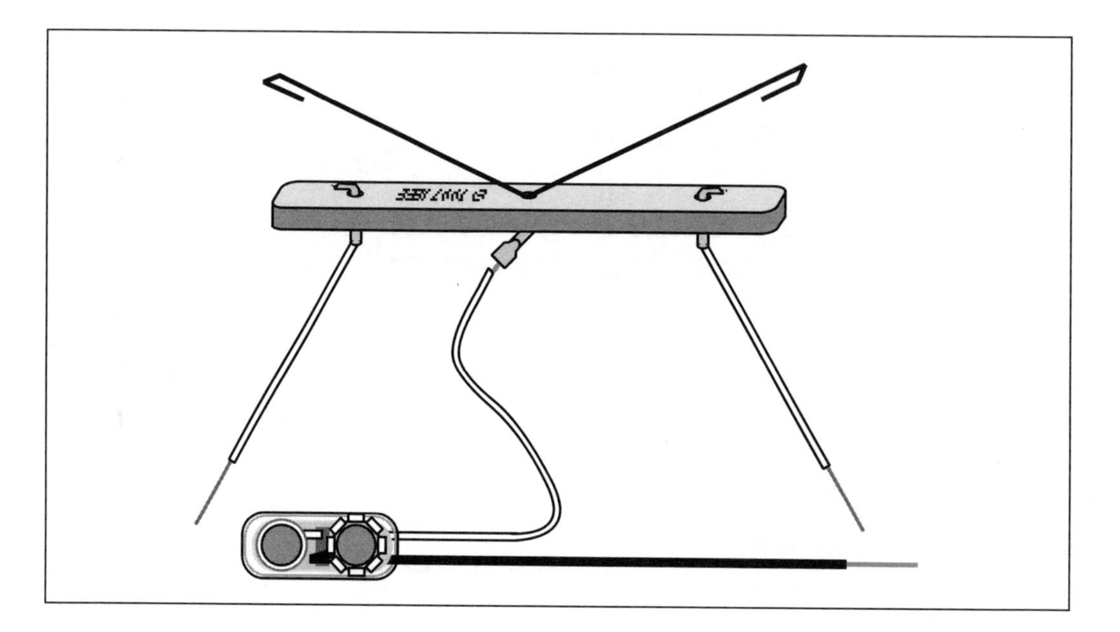

Figure 6.13. Bend the V-clamp ends.

The external arms of the V-clamp form the two manual actuators of the double switch. The bend of the V-clamp inside the handpiece connects the power crimp to the double switch.

This completes the manual controller. See Figure 6.14.

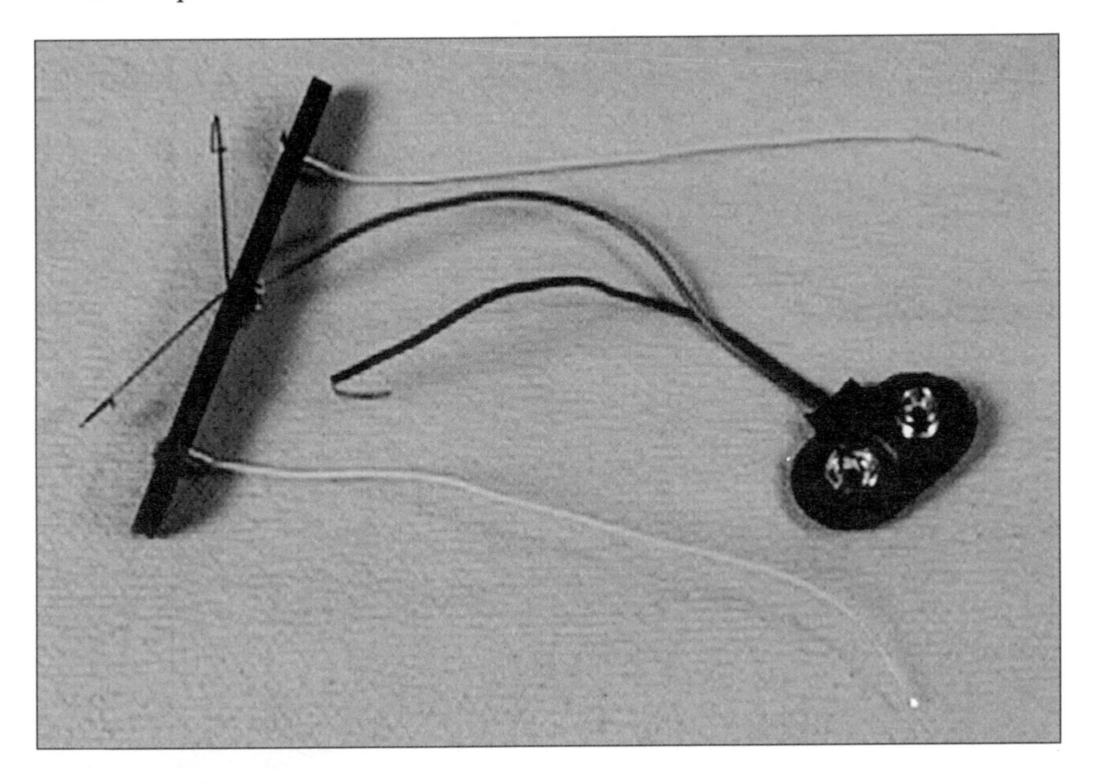

Figure 6.14. Completed manual controller.

ATTACHING THE TETHER BETWEEN STIQUITO AND THE CONTROLLER

To use the manual controller to make Stiquito walk, cut three 500-millimeter (50-centimeter) lengths of 34 AWG magnet wire. Sand the insulation from 14 millimeter of each end of the magnet wire (Figure 6.15).

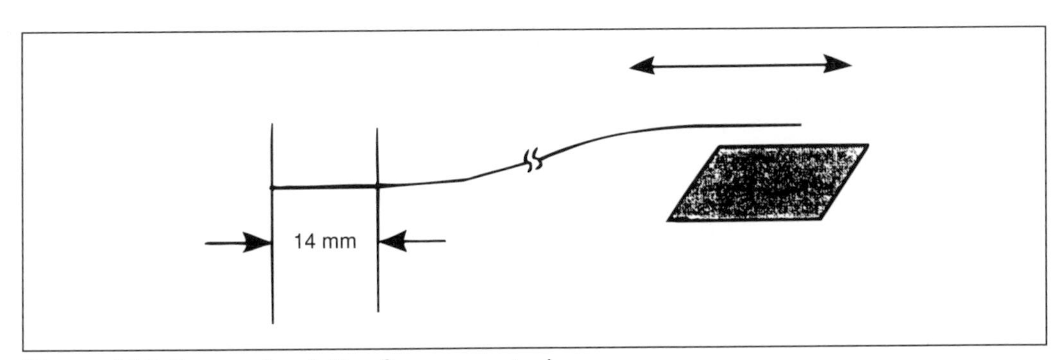

Figure 6.15. Remove insulation from magnet wire.

Use the six connecting crimps to attach the bare magnet wire to the ends of Stiqito's power bus and control wires and to the ends of the manual controller's wires. Attach the wire between the Stiquito power bus and the battery clip first to prevent wiring the tether incorrectly. Attach a 9-volt cell to the terminal, then briefly and alternately press each arm of the double switch. See Figure 6.16. Stiquito should walk with a tripod gait. If it does not, see the next section for troubleshooting suggestions.

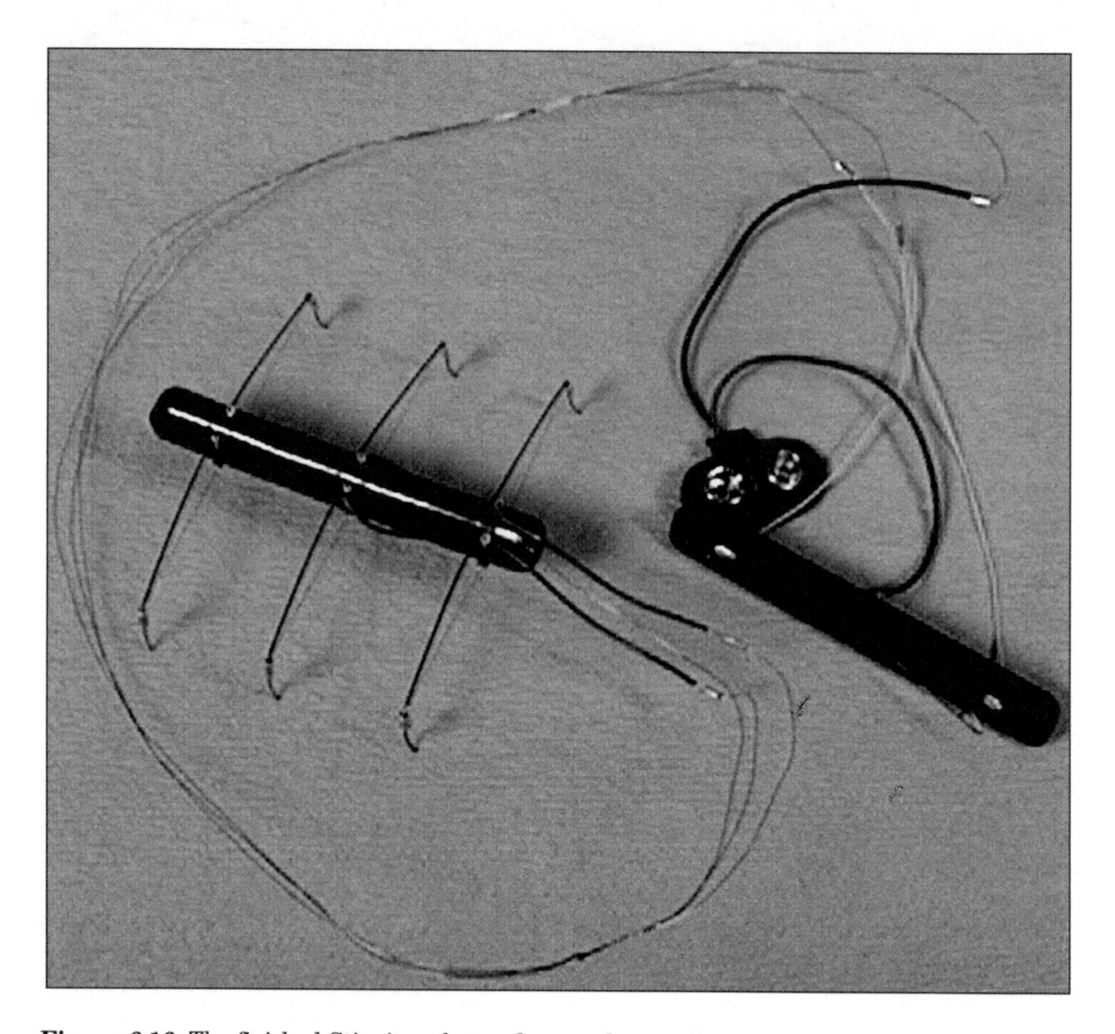

Figure 6.16. The finished Stiquito robot and manual controller.

TROUBLESHOOTING

Troubleshooting is applied logical deduction. To avoid the frustration encountered when a project fails to work as expected, *expect it not to work.* Take this attitude from the start, and think about factors that could affect the operation of the robot and the behavior (or lack of it) that would result. Then, when the inevitable happens, you will have a set of hypotheses about why the robot failed to work. The hypotheses might be wrong, but that is OK: Wrong assumptions lead to right deductions if you are willing to discard assumptions that are not supported by experimentation. The schematic in Figure 6.17 may be a helpful reference during troubleshooting. Refer to Appendix B for suppliers of materials or repair kits should you need them.

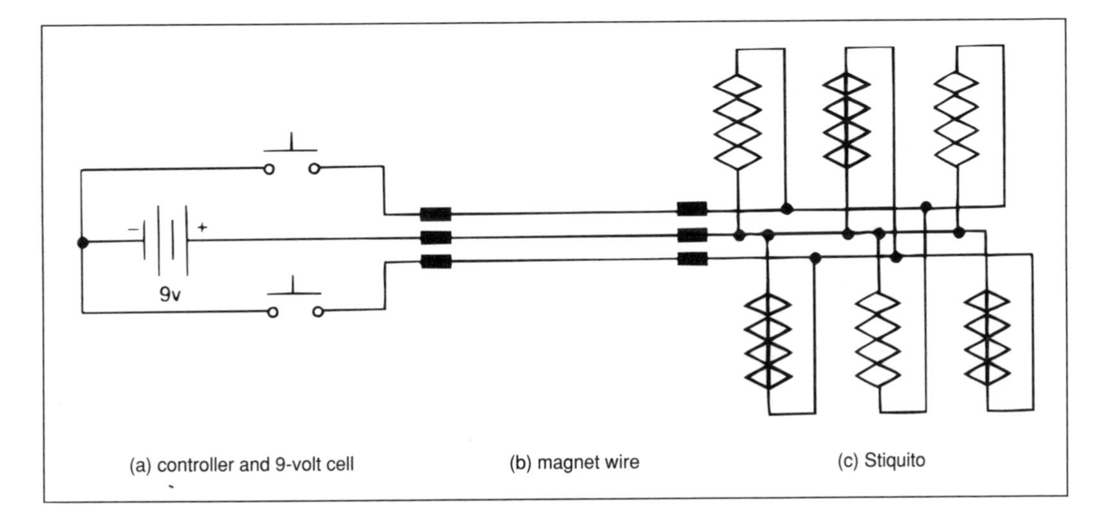

Figure 6.17. Controller, magnet wire, and Stiquito circuit schematic.

Stiquito Walks in a Straight Line—This is perfect. There is no trouble. Congratulations on a good job!

Stiquito Walks to the Left or Right, But Not Straight—This is OK and is very common. The reason is that one or more of the legs is crooked, or does not operate at full extension. Check for loose actuator wires; loose crimps; poor electrical connections (look for open or high-resistance paths from the power supply to the legs and back); a weak power supply (nitinol draws about 180 milliamps per leg, which will drain a 9-volt cell after several weeks of occasional use); legs that are not parallel; or ratchet feet bent at different angles.

Stiquito Does Not Walk at All—If there is nothing obviously broken or loose (actuator wires, crimps, control wiring, magnet wire), then check the power source. The 9-volt cell might be dead. Check the power bus and the body crimps for shorts. Check for shorts between connecting crimps (it might be necessary to hold them apart with a piece of tape). If the ratchet feet are not bent, or if Stiquito is operated on a smooth surface, the legs might move but will not catch the surface—and Stiquito will thrash around but not walk. Stiquito walks best on lightly textured surfaces such as indoor-outdoor carpet, a cloth-covered book, pressboard, or poured concrete. Stiquito walks poorly on glass, smooth plastic, or tiled floors.

Leg Moves to Full Extension (4–5 millimeters)—This is perfect. You obviously built Stiquito carefully.

Leg Does Not Move at All—The probable causes include a very loose actuator wire or a dead 9-volt cell. Check the cell with a voltmeter. If the power is OK, then check the electrical connection between the control wire and the body crimp. Use the voltmeter or apply power to various points in the circuit until the leg moves. Do not use a 9-volt cell to test a single leg; it will snap the actuator wire. Test single actuator wires using 3 volts supplied by two 1.5-volt AA cells in series. If the leg cannot be activated from points farther away in the control circuit, then there is an open circuit between that point and the actuator wire. If the electrical connections are good, examine the actuator wire. If it is loose, or the body or leg crimps are not tight, the actuator is too slack to operate. Tighten the actuator wire (it might be necessary to remove the old leg crimp and attach a new one), then test the leg. It should work.

Leg Moves Slightly (1–3 millimeters)—The actuator is probably loose, but is taut enough to take up the slack and then move the leg. Retension the actuator, replacing the leg crimp. Another cause is increased resistance as the crimps age. Aluminum oxide builds up inside the crimps; this can be alleviated by operating the leg, causing the retaining knots to expand and improve contact with the aluminum inside the crimp. Squeezing the crimp also helps to improve the electrical connection.

Leg Moves in One or More Jerks—There is probably an intermittent open or shorted connection. **If the leg jerks backward continuously** in small increments, **remove power immediately** or the actuator wire might be damaged. Check the manual controller for an intermittent connection between the double switch and the contacts.

Leg Heats Up, Smokes, and/or Melts Plastic Near Body Crimp—A little smoke is typical when the actuator wires are first used, probably as oils or oxide or the nitinol wire burn away. But **if there is a lot of smoke** (equivalent to a cigarette left on the edge of an ashtray), or **if something smells hot, like burning or melting plastic, remove power immediately** or the actuator wire might be damaged. There might be a short in the wiring. You might be powering a slack actuator for longer than one second. A slack actuator will not work. Continuing to apply power to the leg will only cause the nitinol actuator wire to overheat, damaging the wire and heating the body crimp enough to melt the plastic body.

Leg Works Well for a While, Then Movement Stops Altogether—The actuator wire has developed slack, a connection has broken, or the battery is dead. If the actuator wire is slack, it might need to be retensioned and recrimped. Check for broken connections, especially where the magnet wire is crimped. Try a fresh 9-volt cell.

Leg Works Well for a While, Then Movement Diminishes by 2 millimeters or More—The actuator is probably somewhat loose, but is taut enough to take up the slack and then move the leg. Another cause is increased resistance as the crimps age. Apply power to the leg and/or squeeze the body and leg crimps to improve the electrical connection.

Magnet Wires or Control Wires Break—Magnet wire breaks easily. If it does break, remove the connecting crimp, sand away some insulation, and crimp it back into place. If the control wires break, this will probably occur as the hardwired gaits are being changed. Do not bend any wires back and forth repeatedly; the wire will become fatigued and break.

Actuator Wire Breaks—If the actuator wire breaks during assembly, it was sanded too much or was nicked (probably while removing the knot from a music wire leg). Actuator wires can also break from these causes during operation. Do not power a single actuator wire with a 9-volt cell; the wire will contract so rapidly that it cannot overcome the inertia of the leg, and the actuator wire will snap.

Other Controllers

After you have mastered making Stiquito walk with the manual controller, you are ready to use an analog or digital controller. See Chapters 7 and 8, or the first Stiquito book,[1] for ideas on how to further control Stiquito.

EXERCISES

1. Why do you think you need to sand the insulation off the magnet wire?
2. Does it matter which battery clip wire you attach to the Stiquito power bus (the black or the red)? Why or why not?
3. **How long will a 9-volt battery last?** To understand how long you can run Stiquito with a 9-volt battery, you need to consider how much current is stored in the battery and how much current Stiquito uses.

   A battery's storage capacity is stated in milliamp hours. For example, a 200 milliamp hour battery supplies 200 milliamps of current for one hour or 100 milliamps of current for two hours. Assume that, based on your pressing the switches, three Stiquito legs are active for one second, then all legs are at rest for one second, then the other three legs are active for one second, then all legs are at rest again for one second. Each leg uses 180 milliamps. How long will it take to drain a 200 milliamp hour battery?

BIBLIOGRAPHY

1. Conrad, J. M., and J. W. Mills. 1997. *Stiquito: Advanced experiments with a simple and inexpensive robot,* Los Alamitos: Calif.: IEEE Computer Society Press.

Chapter 7

A PC-Based Controller for the Stiquito Robot

James M. Conrad

One of the most flexible ways to control the gait of a Stiquito robot is by using a personal computer (PC) and writing a program. The program, when run, will control the contractions of the nitinol wire, thus making Stiquito walk. This chapter contains the instructions for making a circuit that can be plugged into the parallel port of a PC. This chapter also discusses the concepts of the PC parallel port, contains instructions on how to build the board circuitry, and provides instructions on how to write a program to make Stiquito walk with a simple tripod gate.

Although you can build several circuits to attach to the parallel port, we recommend using a circuit that does not draw current from the PC. Several kits may be available, but these instructions show how to build the circuit board and program the Stiquito to walk using the Stiquito parallel port kit available from some of the suppliers listed in Appendix B.

THE BIG PICTURE—WHAT DOES THE CIRCUIT DO?

The printed circuit board plugs into a personal computer parallel port. It can generate enough current to light up LEDs on the board and make Stiquito walk. The LEDs are used to help you develop your computer program. This board provides an easy-to-see report on how your program is executing. Once your program works correctly and the LEDs show a correctly working gait, you can plug the Stiquito control wires into a socket on the board.

The schematic in Figure 7.1 shows the logic on the circuit board when Stiquito is not attached. The ULN2803 driver chip will invert the value of the input. In that way the LED will light up when current is drawn toward the ULN2803. Figure 7.2 is the same as Figure 7.1 but with the Stiquito robot attached. In this circuit, the LEDs will light and the nitinol legs will contract.

The circuit shown in Figure 7.1 should be used without Stiquito attached to test your hardware and software. This will prevent Stiquito's nitinol actuators from damage while you are developing your circuit. Figure 7.2 shows the equivalent circuit once Stiquito is attached.

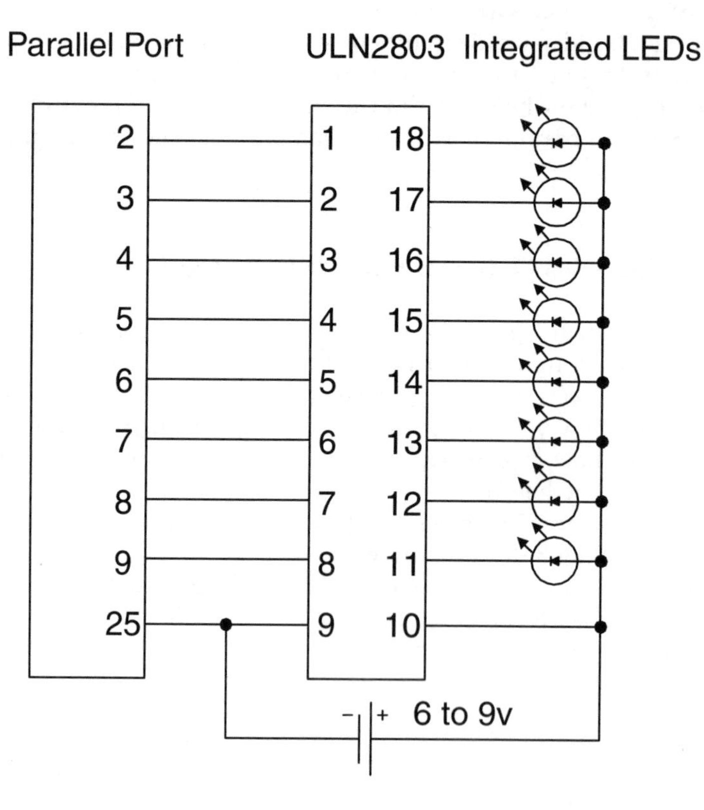

Figure 7.1. An electronics schematic for the PC parallel port controller.

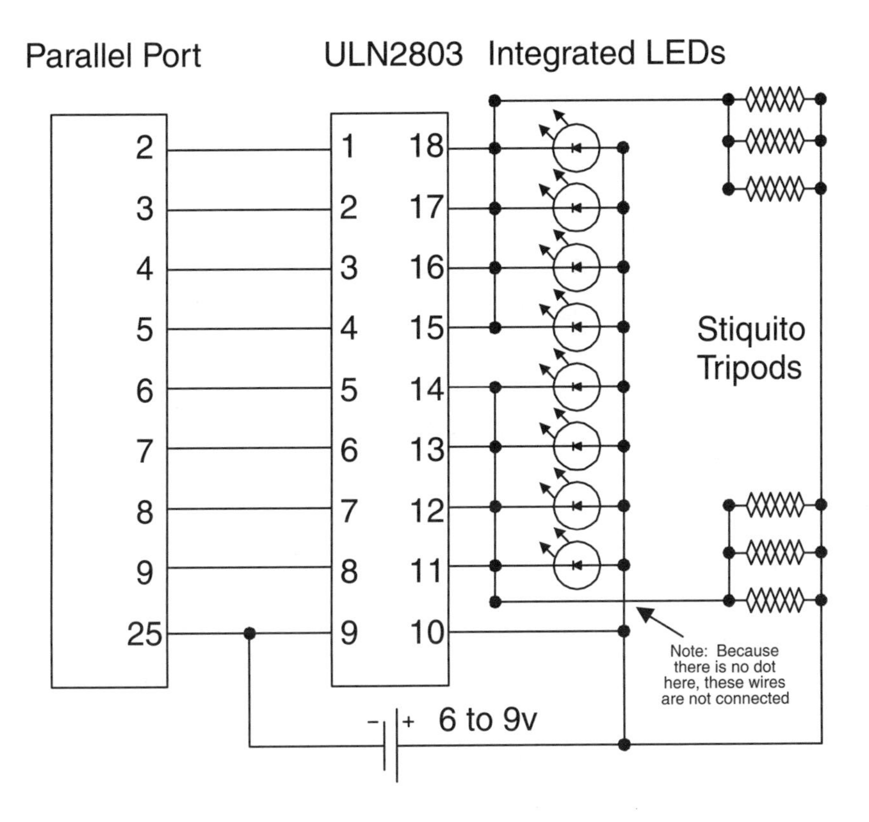

Figure 7.2. The Figure 7.1 schematic with Stiquito attached.

MORE TOOLS AND RULES

Because you must solder the electronic components to the card, additional tools are needed to build the interface. The tools **required** are:

- Needle-nose pliers
- Wire cutters
- Small hobby knife (X-Acto™ type)
- Soldering iron
- Solder
- Volt-ohmmeter
- 320-grit fine sandpaper
- Electrical tape

You will also need to buy a 9-volt battery to power your robot. Or you can use any 5-volt to 9-volt power source (battery or DC power supply capable of providing up to 1,000 milliamps of current).

Look for these cues (small pictures) as you follow the instructions to assemble the interface card.

Cue	Meaning
✓	**Do** take the action indicated. No particular precision is required.
✗	**Don't** do this! It is a mistake.
✍	**Solder the part** at the point or points indicated.
⌶	**Some variation is allowed.** The dimensions are not critical. Stay within 1 or 2 mm of the specified measurement.
◉	**A loose or sloppy fit is OK.** Your interface will work fine if the fit is not tight.

PRECAUTIONS

- ✓ Soldering irons get **very hot.**
- ✗ Do **not** touch exposed metal on a hot soldering iron. Hold it by the insulated handle.
- ✗ Do **not** lay a hot soldering iron on the work surface or flammable material.
- ✓ Use a soldering iron holder or lay a piece of wood under but not touching the tip to protect your work surface.
- ✗ Although the integrated circuits used in the interface are not particularly static sensitive, do **not** handle them by the pins.
- ✓ In winter and dry weather, touch a large metal object (table, door frame) to discharge any static electricity before handling integrated circuits.
- ✓ ALWAYS wear safety glasses when performing this work.

Now let's build the interface card and use it to control your robot.

IDENTIFY PARTS

Make sure you have all the parts. Match them against the photograph in Figure 7.3 to be sure they are the correct parts. Please note that some supplied parts may not look *exactly* like those in the figures, but are very close. For example, the LEDs may be larger, or the circuit board may have more holes.

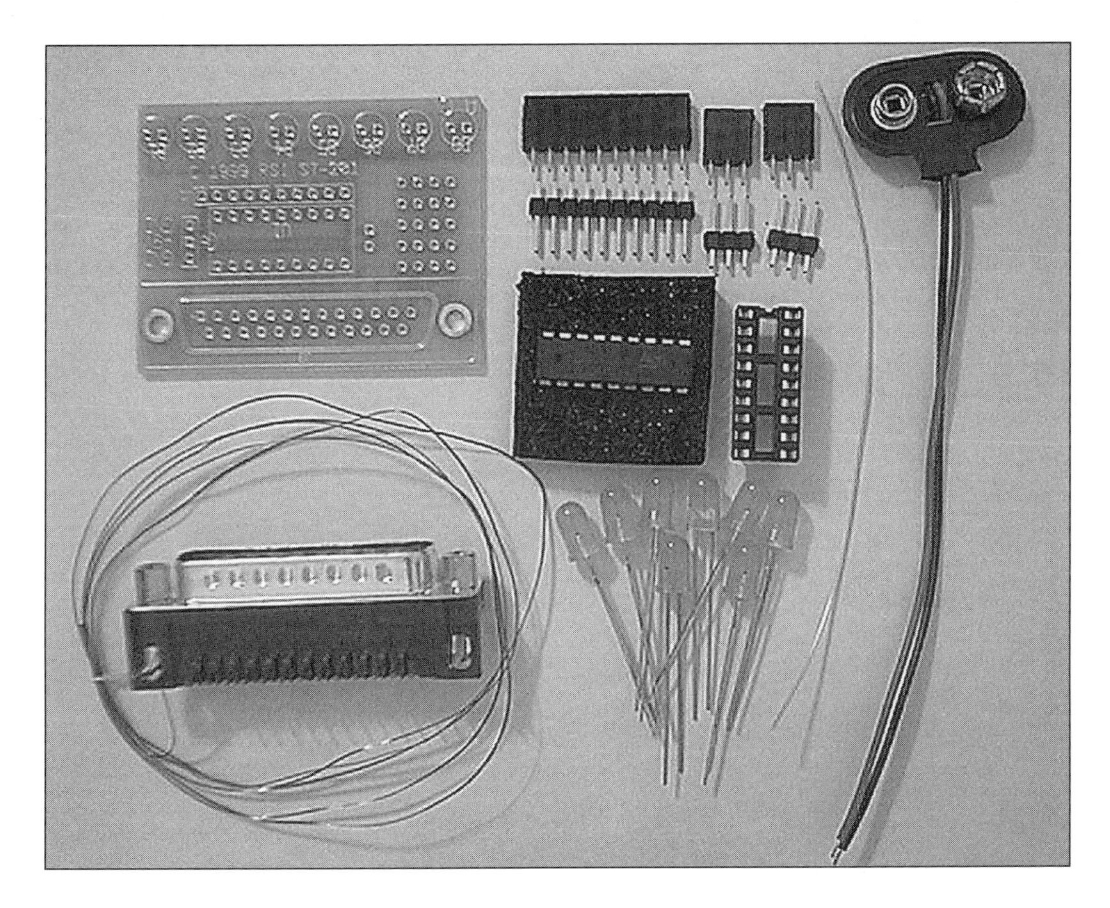

Figure 7.3. Parts used to build the PC controller.

Qty	Name	Where Used or Location on Board
8	Integrated resistor lamp (LED)	board-D1 through D8
1	8-transistor Darlington Array chip (ULN2803)	board-U1a
1	18-pin DIP socket	board-U1
2	3-pin socket	board-J2 and tether
2	3-pin jumper post	battery and Stiquito
1	10-pin socket	board-J1
1	10-pin jumper post	tether
1	9V battery connector (optional)	battery
1	25-pin D-shell male connector (circuit board mount)	board-J3
1	150 cm of 3-strand 34 AWG magnet wire, or three 150 cm lengths of single-strand 34 AWG magnet wire	tether
1	10 cm of 28 AWG wire wrap wire	Stiquito
1	Printed circuit board (component side) ST-201	board

SOLDERING SKILLS

Soldering mechanically and electrically connects components and integrated circuits to the interface card. The basic technique for soldering is described here.[1] You may want to practice on an old circuit board before you solder the parts to the interface card.

✓ Use the soldering iron's tip to heat the **pad,** not the integrated circuit pin. When the pad is hot, touch the solder to the heated pad, and the solder will flow onto the pad and the pin.

✓ Use just enough solder to wet the pin and cover the pad.

✓ Each solder joint should be bright, shiny, and have flowed evenly around the pin on the pad. The solder on adjacent pads must not touch.

✗ A solder joint should **not** be dull, cracked, or beaded up on the pad.

✗ A solder joint must **not** cross between two pads, or a pad and a trace. This will create a short circuit. Your interface card will almost certainly **not work correctly.**

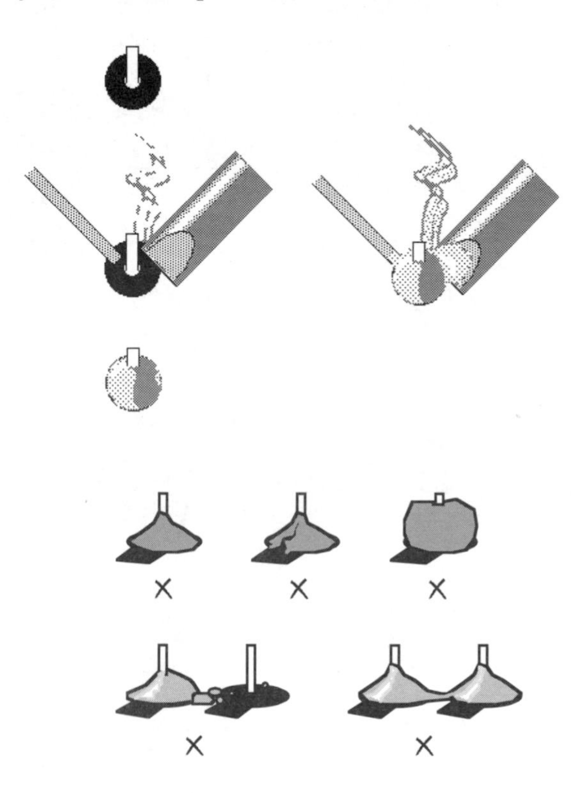

After you have finished soldering, you should examine the board for workmanship errors. Check the board for short circuits and broken traces:

✓ Examine the wiring side of the interface board. Look at places where one trace (wires on board) or pad (round circle on board) is near another; check that they do not touch. Look at long traces and near bends; check that the trace is not broken at that point.

✓ If traces or pads touch, but they should not, use the knife to cut the unwanted connection.

✎ If a trace is broken, lightly sand it on either side of the trace, then solder the broken ends together using a piece of fine wire to bridge the gap.

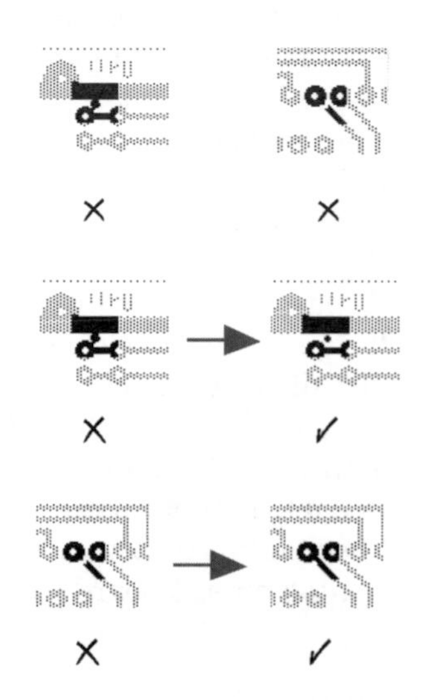

MAKING THE PRINTED CIRCUIT BOARDS

The steps involved in making the parallel port board involve inserting the sockets, LEDs, and connector into the board (Figure 7.4) and soldering.

✓ Take the PC parallel port printed circuit board and turn it so the square markings are face-up. These markings will tell you where parts are to be inserted.

◉ Take the 18-pin socket and insert it into the board location labeled U1. Make sure the notch at the top edge of the socket matches the notch in the white drawing on the board.

✓ Holding both the socket and the board, turn the board over and lay it flat.

✓ Slightly bend two pins at the corner of the socket toward the outside of the socket. Once you have bent the pins, the socket should be held in place on the board. Make sure not to bend the pins too much (a 30 degree angle should be fine).

✓ Turn the circuit board back over and repeat the process for the 10- and 3-pin header sockets for locations J1 and J2, respectively.

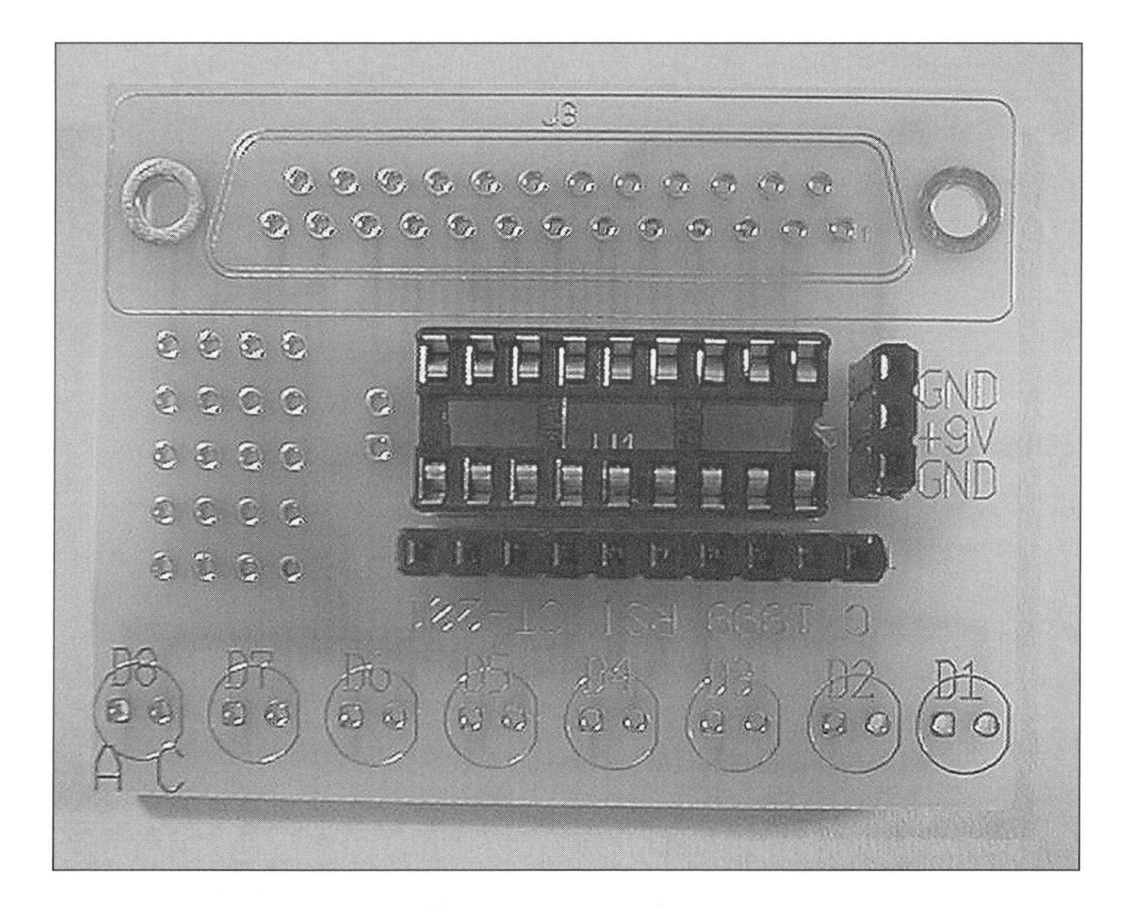

Figure 7.4. Inserting connectors and sockets into the circuit board.

✓ Turn the printed circuit board so that the electrical traces are face-up.

✍ Use your soldering iron and solder to attach the connectors to the board (Figure 7.5). Follow the soldering instructions discussed earlier in this chapter.

✓ Make sure to inspect your soldering work.

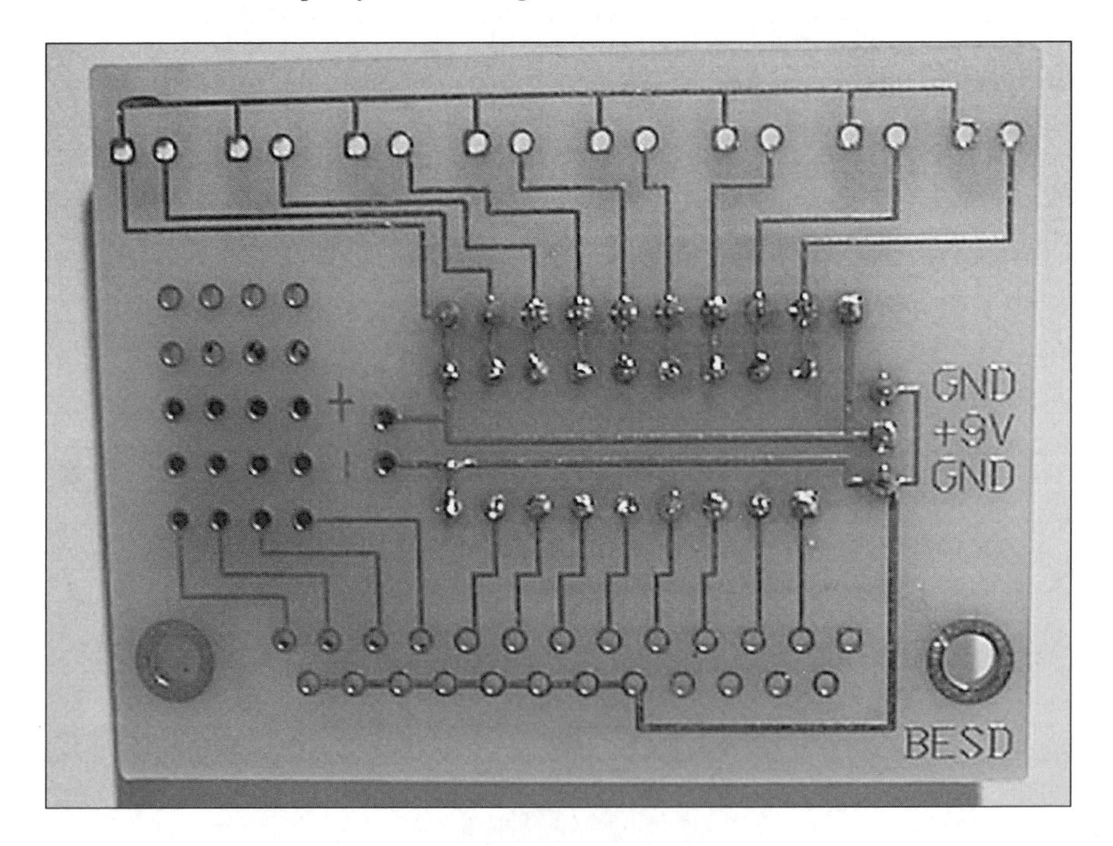

Figure 7.5. Solder the connectors and socket.

◉ Turn the printed circuit board (PCB) component side-up and insert the 25-pin connector into the board.

✓ Turn the PCB solder side-up.

✍ Solder the pins of the connector to the board (Figure 7.6).

✓ Again, inspect your work when you are done.

Chapter 3 discussed the polarity of LEDs. Make sure you insert the LEDs into the board in the correct orientation.

✓ Turn the PCB component side-up.

✓ Locate the location for the LEDs. These are labeled D1 through D8. "D" stands for "diode," since this device is a light-emitting diode. Notice that the markings on the board have a flat side. Ensure that when you insert the LED into the board, you put the flat side of the LED on the flat side of the markings. See Figure 7.7.

Figure 7.6. Solder the 25-pin connector to the circuit board.

Figure 7.7. Placing the LED onto the circuit board.

◉ Insert one LED.

✎ Solder these two wires to the PCB.

◉ Next, take your wire snips and cut the LED wires above the solder joint.

✓ Repeat this process for the other seven LEDs.
✓ Turn the PCB component side-up.

◉ Carefully insert the ULN3903 chip into the 18-pin socket. Be careful not to bend the pins (Figure 7.8).

✓ You may find the chip pins are slightly wider than the socket. One trick to bend all of the pins closer to each other is to lay the chip on its side and gently press down on the other side until the pins bend slightly (see Chapter 8, Figure 8.8). This step may take a little practice.

Figure 7.8. Inserting the integrated circuit into the circuit board socket.

Your PC parallel printed circuit board is now complete.

Using your ohm meter, put one contact on the pin labeled J1 of the 10-pin header, and the other contact on each of the other eight pins, one at a time. Ensure that the ohmmeter registers some resistance, but not infinite resistance. Check your work to make sure you have no shorts or broken traces.

✎ If you plan to use a 9-volt battery to power your board, you need to solder the battery connector to the 3-pin jumper post (Figure 7.9). Solder the black wire to one of the outside pins. Solder the red wire to the center pin.

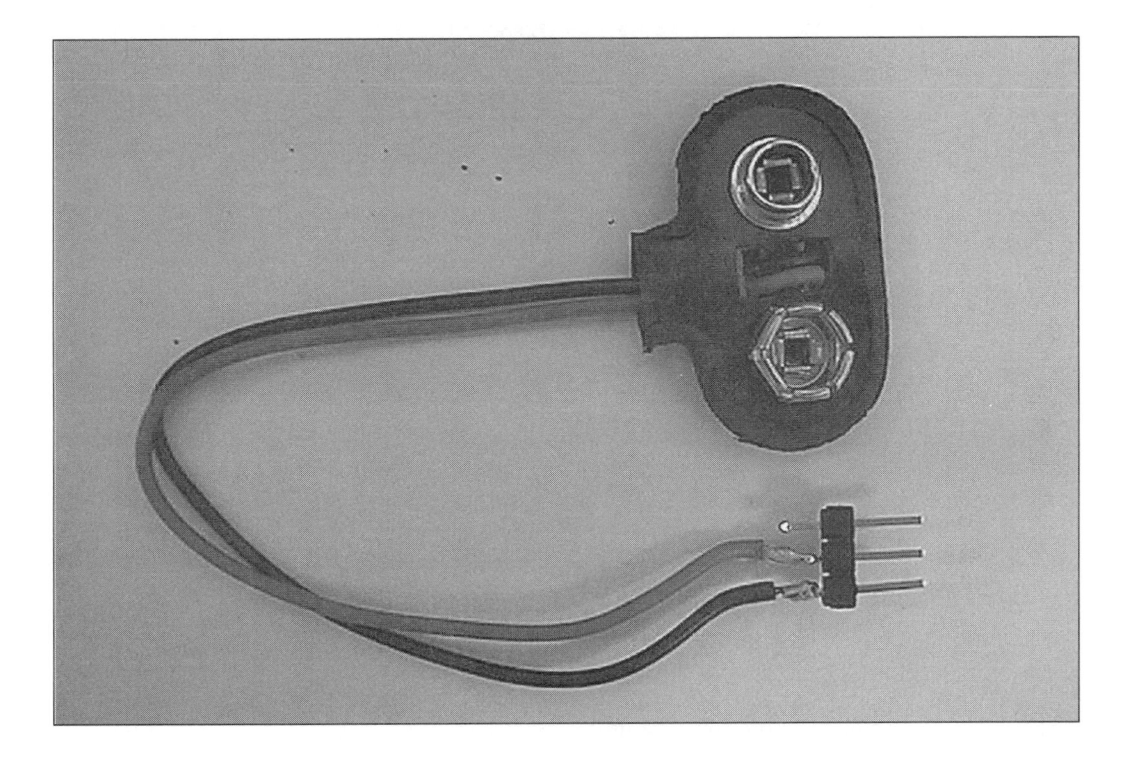

Figure 7.9. Soldering the battery connector leads to the jumper post.

ATTACHING STIQUITO TO THE PARALLEL PORT CIRCUIT BOARD

Now that you have built the parallel port controller board, you need to prepare the Stiquito robot and make its control tether.

- Cut a small length of wire wrap wire.
- Solder it to the center Stiquito Bus bar, then to the center 3-pin jumper. See Figure 7.10.

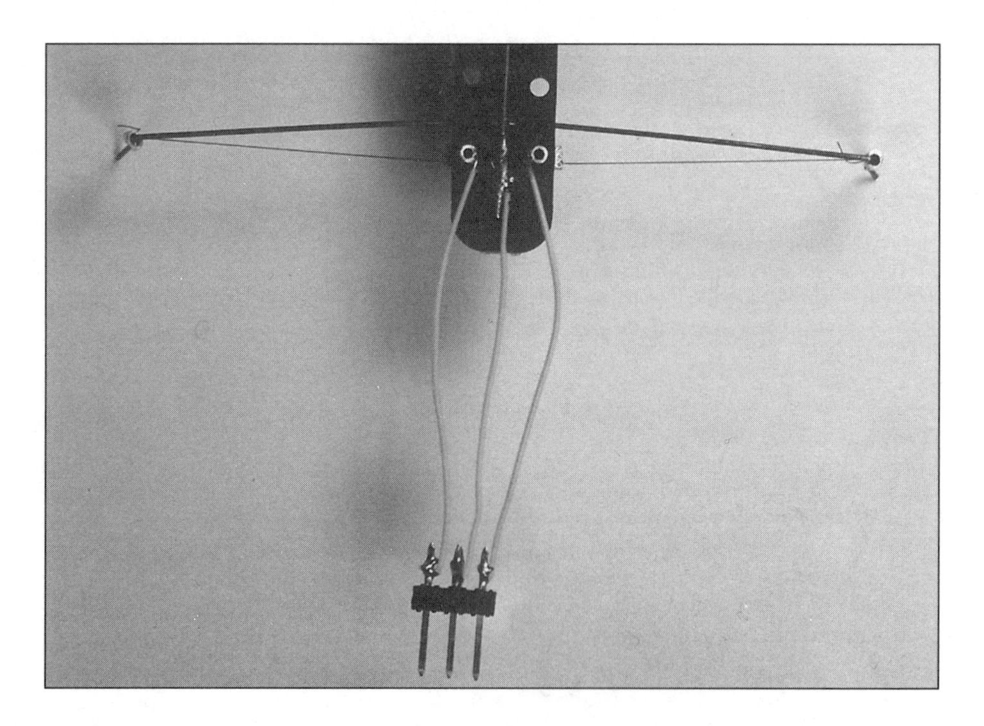

Figure 7.10. Soldering a connector to Stiquito.

- Solder the other two Stiquito tripod control wires to the two outside pins of the 3-pin jumper.
- ✓ Sand all six ends of the three wires of the magnet wire group.
- Solder the three wires at one end of the magnet wire to the three pins of the 3-pin socket.
- Identify the wire soldered to the center of the 3-pin socket. Solder it to the first pin of the 10-pin jumper post.
- Solder one of the remaining wires of the tether to the next four pins of the 10-pin socket. Use some of the wire wrap wire to connect these four pins together.
- Solder the remaining wire of the tether to the next four pins of the 10-pin socket. Use some of the wire wrap wire to connect these four pins together. This should leave one pin of the jumper empty.

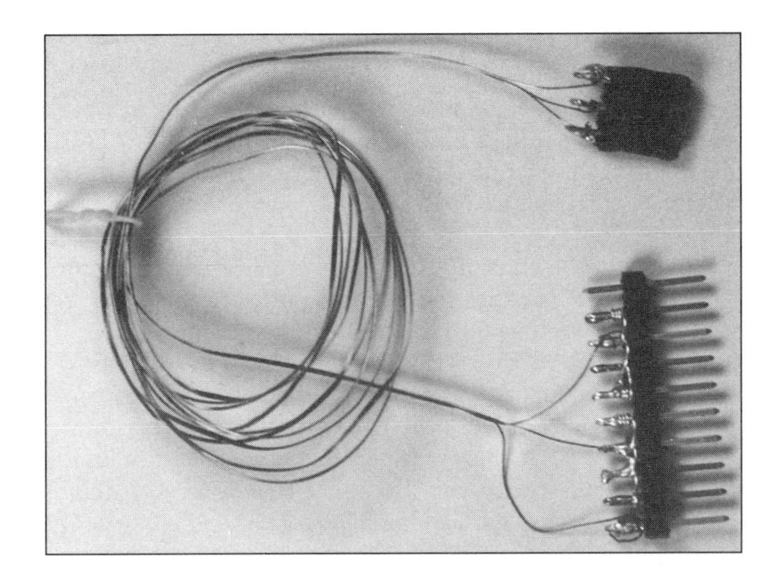

Figure 7.11. Soldering the tether.

Plug the tether into Stiquito, but do not plug the tether into the parallel port card yet (see Figure 7.12). You will use the board to test your Stiquito walking program by observing the LEDs before making the legs move (and perhaps preventing damage to the wires due to a programming error).

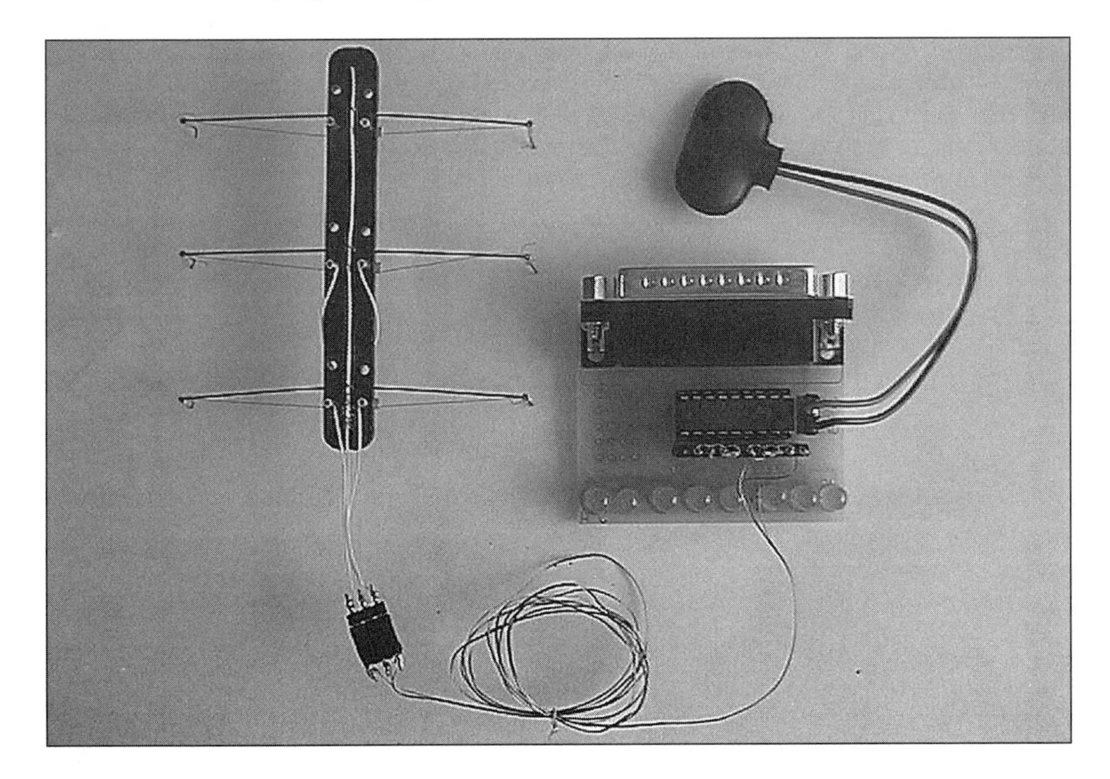

Figure 7.12. The completed kit with Stiquito attached.

THE PERSONAL COMPUTER PARALLEL PORT

The parallel port was designed to serve as an output port from a personal computer. It was designed to attach to a printer. Some parallel ports allow both input and output. We only use the port for output. Although there are 25 pins for a parallel port, we only use nine. Eight of the lines are used as data output lines, and one line serves as the electrical ground.

When using the parallel port, computer programmers usually write information to two locations. One "register" location controls the port and the other "register" location contains the data to send. We use only the data register.

Some computers have more than one parallel port. These ports would be labeled LPT1, LPT2, and LPT3. Each port has a different data register, so you can access each port by using different register addresses. Typically the register address for the single parallel port (or LPT1) is &H378, but your particular PC may use another address like &H278 or &H3BC. You can verify this by using the Microsoft diagnostics program (MSD.EXE) and examining the "port address."

We will use all eight of the parallel port output lines to control our Stiquito robot. We will control each line with a binary digit, or **bit.** A "1" means turn on the line and a "0" means turn off the line. To write to the parallel port, we will write eight bits of data to the parallel port's data register. For example, to write the signal "1" to the top four bits and "0" to the lower four bits of the register, we would send the eight bits "11110000" to the port. This would be written in the QBASIC programming language as: `OUT &H378, data` where *data* is the bit pattern "11110000." A grouping of eight bits is known as a **byte.**

We cannot represent the bit pattern "11110000" as data in the QBASIC language, but we can convert it to **hexadecimal** representation. In hexadecimal, numbers are made by grouping four bits together and converting them to the numbers 0 to 9 and letters A to F. This grouping of four bits is called a **nibble.** Yes, that's right, two nibbles make one byte. Table 7.1 shows the representation of a nibble in both bits and hexadecimal, and their equivalent decimal value.

Table 7.1 Hexadecimal Representation

Bit Pattern (nibble)	Hexadecimal	Decimal
0000	0	0
0001	1	1
0010	2	2
0011	3	3
0100	4	4
0101	5	5
0110	6	6
0111	7	7
1000	8	8
1001	9	9
1010	A	10
1011	B	11
1100	C	12
1101	D	13
1110	E	14
1111	F	15

For our Stiquito control application, we will only use nibble values of "0000" (0) and "1111" (F). To define the value a hexadecimal number, we put &H in front of the digits. The line above is now written: `OUT &H378, &HF0`

PROGRAMMING A GAIT FOR STIQUITO

The mechanisms of arthropod locomotion are complex and have been extensively studied. The structure of an insect leg is also quite complicated. But, even though Stiquito is simple, the fundamental features of arthropod locomotion can be demonstrated by small programs. Later, if you choose, you can develop more realistic models of gait controllers based on neural networks or central pattern generators, and feedback from strain gauges or other sensors that mimic the sensorimotor loop in a real insect.

The gaits of insects are believed to be due to central pattern generators that vary the animal's gait from a **metachronal wave** to a **tripod gait,** and all the variations in between. Each gait conserves energy as it preserves the balance of the insect. As the animation sequences in Figure 7.13 indicate, the insect is always in a stable position with at least three legs, and often more, on the ground at all times.

The metachronal wave is the slowest and most stable gait. It is seen when a "wave" of leg movement ripples down each side of the insect or arthropod. The animation sequence shows two "waves" flowing down each side of a ten-legged insect robot.

The tripod gait is the fastest and most stable gait, with two legs on one side of the insect and one on the other side alternately on the ground or in the air, as shown in Figure 13. This tripod gait shown relies on a leg that has two degrees of freedom, like Stiquito II[2]. The Stiquito assembled using this book has only one degree of freedom.

We will make Stiquito walk with a simpler form of the tripod gait shown in the example. We will only flex and relax the legs while they are on the ground, the same way it is controlled using the manual controller from Chapter 6.

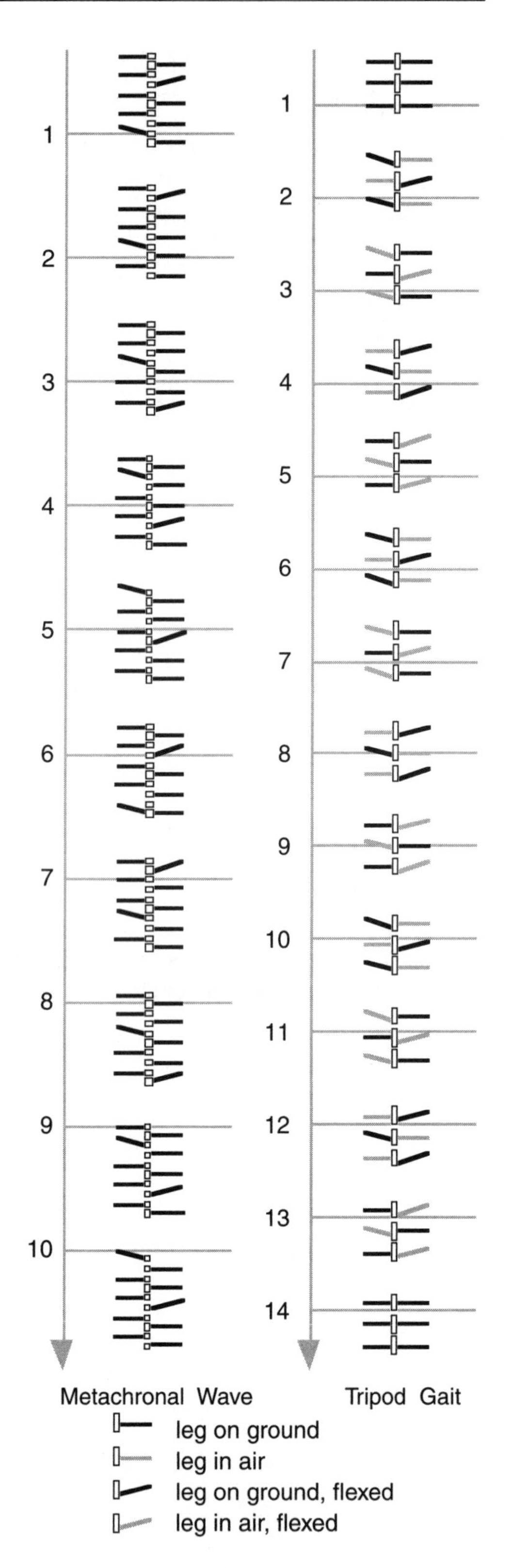

Figure 7.13. Insect Gaits.

Using QBASIC to control the walk, you will need to use the OUT statement to activate, then deactivate the legs. You should add a delay in your program to hold the activation signal for about 1 second, then hold the deactivation signal for 1 second. The code to perform this task for one tripod is:

```
REM Program to make Stiquito walk with a tripod gait. This assumes
REM that the upper nibble controls one tripod, and the lower nibble
REM controls the other. We allow the nitinol to rest after it is
REM activated.
   REM "OUT &H378" sends an 8-bit value to the printer port. The
   REM data sent is hexadecimal.

   DELAY = 14000
 10 OUT &H378, &HF0 : REM &HF0 is binary 11110000
   FOR x = 1 TO DELAY : NEXT x

   OUT &H378, 0
   FOR x = 1 TO DELAY : NEXT x

   OUT &H378, &HF : REM &HF0 is binary 11110000
   FOR x = 1 TO DELAY : NEXT x

   OUT &H378, 0
   FOR x = 1 TO DELAY : NEXT x

   REM If a key on the keyboard was pressed, then end.
   REM Otherwise, walk some more!
   a$ =INKEY$
   IF a$ = "" THEN GOTO 10
   END
```

The number 14000 is an arbitrary value that is computer dependent. You may have to make this number higher if your computer is faster than an 80486-based machine (66 MHz).

SAVING POWER BY PULSING THE CONTROL SIGNAL

Driving the nitinol actuator with the same amount of current is unnecessary after the nitinol has contracted. Only enough current to keep the nitinol contracted is needed, which is just enough to replace the energy that escapes as heat. The current and the voltage supplied to the nitinol cannot be changed dynamically, but the power can be varied using a technique called **pulse frequency modulation (PFM).**

Pulse frequency modulation means that the number of pulses (their frequency) is varied over time. The PC parallel printer port and the interface card can generate a PFM signal because the nitinol reacts slowly, compared to the speed with which a BASIC program can turn the ULN2803 driver chip on and off. By varying the length of time that the driver chips are left off, the frequency of the pulses can be increased or decreased. This allows the power used to drive the robot to be varied dynamically.

The nitinol actuator responds to the heat generated by the current pulses, and the loss of heat due to convection from the wire. Figure 7.14 shows how a PFM driver program works.

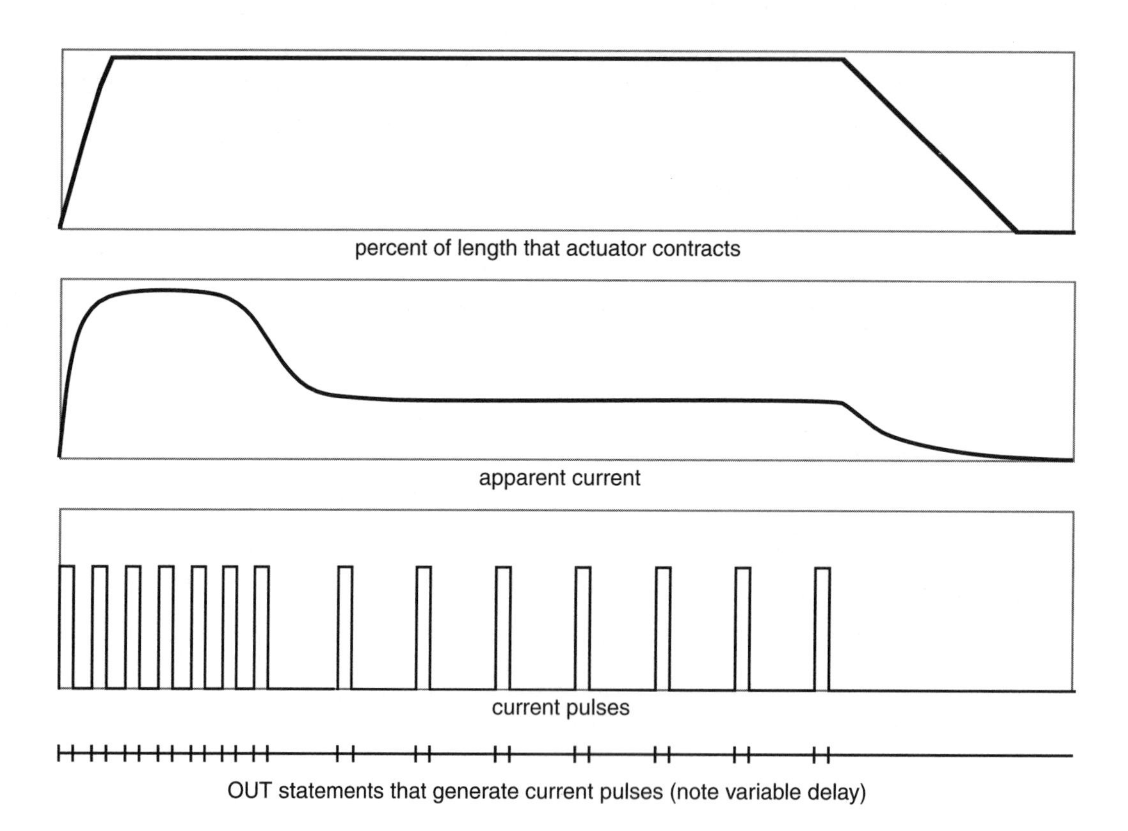

Figure 7.14. Pulse frequency modulation.

The following code shows how to use PFM to control one tripod of the Stiquito robot:

```
REM High frequency pulses initially contract actuators
FOR a = 1 TO 20
  OUT &H378, &HF0                 : REM &HF0 is binary 11110000
  FOR x = 1 TO 100 : NEXT x
  OUT &H378, 0
  FOR x = 1 TO 100 : NEXT x
NEXT a

REM Low frequency pulses maintain actuator contraction
FOR a = 1 TO 40
  OUT &H378, &HF0                 : REM &HF0 is binary 11110000
  FOR x = 1 TO 100 : NEXT x
  OUT &H378, 0
  FOR x = 1 TO 200 : NEXT x
NEXT a
```

EXERCISES

1. Write a program to make each or the LEDs on the PC parallel port board light up for 1 second, one at a time.
2. Write a program in another language (like C, C++, Assembly, or Java) to make the Stiquito robot walk.
3. To make Stiquito walk as fast as possible, you will want to be able to change the timings of the activations and rests. This means you will want to change the values of the delays in the code. You could also change the programs shown above to allow the user to type a value for the delays using a menu instead of changing the code for each new delay value. Write a program for making Stiquito walk. Allow delays to be entered from the keyboard while the program is running (i.e., from a menu).
4. Write the full program that uses pulse frequency modulation to make Stiquito walk.
5. (Difficult) Create the ideal gait for Stiquito. This requires one control wire for each leg, and thus requires Stiquito be built differently from the instructions included in this book.
6. (Difficult) Build a two-degrees of freedom Stiquito. Create the ideal gait for this Stiquito. This requires one control wire for each forward/back tripod and each lifting tripod. Thus, Stiquito should be built differently from the instructions included in this book.

BIBLIOGRAPHY

1. Mims, F. M. 1988. *Engineer's mini-notebook: Schematic symbols, device packages, design and testing.* Fort Worth, Texas: Radio Shack.
2. Conrad, J. M., and J. W. Mills. 1997. *Stiquito: Advanced experiments with a simple and inexpensive robot.* Los Alamitos, Calif.: IEEE Computer Society Press.

Chapter 8

A Simple Circuit to Make Stiquito Walk on Its Own

James M. Conrad

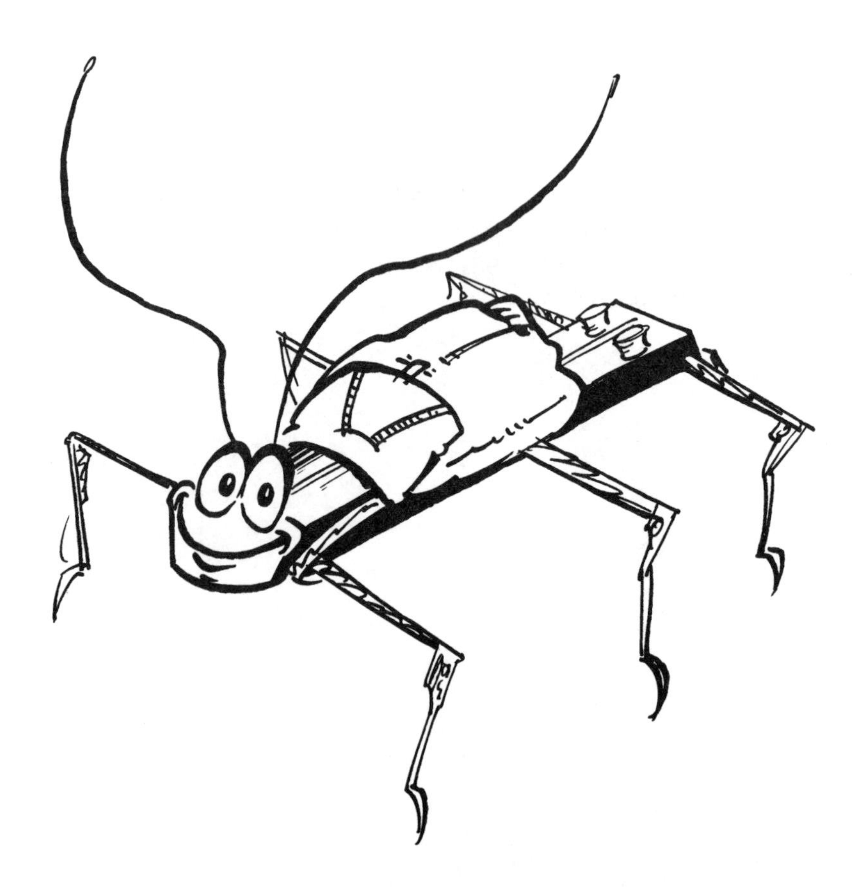

You can build a circuit board that mounts on top of the Stiquito robot and directs the robot to walk in a tripod gait. This circuit will feed current to the nitinol on a periodic basis, which you can adjust. This setup allows Stiquito to be an autonomous robot. A kit containing all of the supplies you need to build this controller can be purchased from one of the suppliers listed in Appendix B.

The card will mount to the top of the already built and tested Stiquito robot. The manual controller should not be attached to the robot.

THE BIG PICTURE—WHAT DOES THE CIRCUIT DO?

The controller circuit described by the schematic in Figure 8.1 generates enough current to contract three nitinol actuators alternately at a time. It uses a popular timer circuit to generate a pulse to contract the legs. The two main parts of the circuit are:

1. The actuator components (LEDs, transistors, and nitinol, for user feedback and motion)
2. The pulsing components (555 timer, capacitors, potentiometer, and resistors for generating the pulses to the nitinol legs)

The potentiometer will allow you to fine-tune the gait to work at an optimal speed.

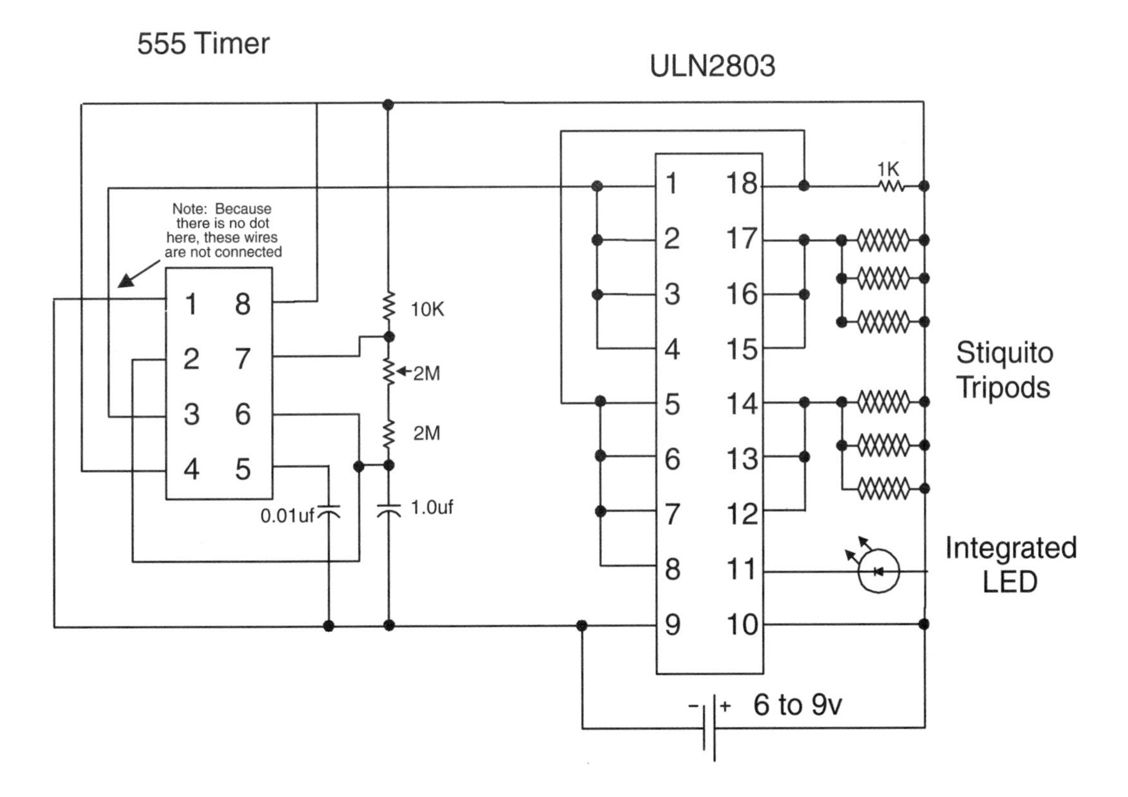

Figure 8.1. An electronics schematic for the analog controller circuit.

THE 555 TIMER CIRCUIT

The 555 timer is a class of timing circuits used to generate a known, periodic pulse. With the addition of a capacitor and two external resistors the 555 can be configured to produce this pulse without any external triggering pulse. An example of this *astable multivibrator* is shown in Figure 8.1 and will be used in this chapter.

The circuitry inside the 555 works based on the voltage sensed at pin 6 and 7. Suppose at time t=0 the output of pin 3 is high (supply voltage = Vcc) and pin 6 is low (ground voltage). The external circuit shown in Figure 8.1 will charge the 1.0 μf capacitor. When the voltage of pin 6 reaches ⅔ × Vcc, the output voltage of pin 3 will switch to 0 volts, and the 555 internal circuitry will start to discharge the 1.0 μf capacitor. When the voltage of pin 6 falls to ⅓ × Vcc, the cycle starts all over again. A timing diagram of the stable operation of this circuit is shown in Figure 8.2.

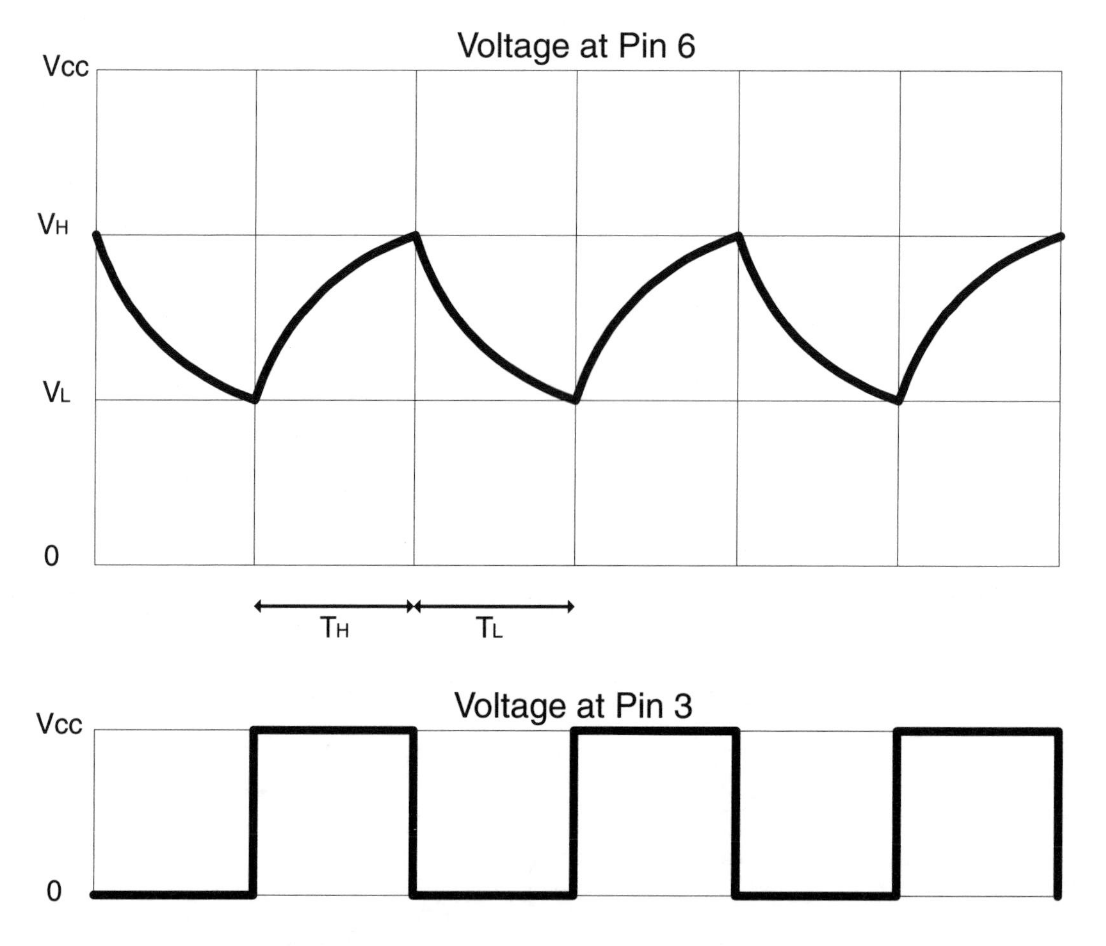

Figure 8.2. The timing diagram of the astable multivibrator, 555 timer circuit.

The times of the high and low outputs on pin 3 can be determined by using similar formulas shown in Chapter 3. We define the following symbols as:

T_L = The time of the 0 volts part of the pulse (low voltage on pin 3)
T_H = The time of the Vcc volts part of the pulse (high voltage on pin 3)
P1 = the value, in ohms, of the Potentiometer (we will assume 0.5 Mohm)
R1 = the value, in ohms, of the R1 resistor (2.0 Mohm)
R2 = the value, in ohms, of the R2 resistor (10 Kohm = 0.01 Mohm)
C1 = the value, in microfarads, of the capacitor connected to pin 6 (1.0 µf)
ln2 = Natural logarithm of 2 = 0.693

The formulas are:

$$T_L = (P1 + R1) * C1*\ln 2$$
$$= (0.5 + 2.0 \text{ Mohm}) * 1.0 \text{ µf} * 0.693$$
$$= 1.733 \text{ seconds}$$
$$T_H = (P1 + R1 + R2) * C1 * \ln 2$$
$$= (0.5 + 2.0 + 0.01 \text{ Mohm}) * 1.0 \text{ µf} * 0.693$$
$$= 1.739 \text{ seconds}$$

The obvious observation from these equations is that if R1 >> R2, then T_L will be close to T_H.

MORE TOOLS AND RULES

Additional tools are needed to build the interface because you must solder the electronic components to the card. The tools **required** are:

- Needle-nose pliers
- Wire cutters
- Small hobby knife (X-Acto™ type)
- Soldering iron
- Solder
- Volt-ohmmeter
- 320-grit fine sandpaper
- Electrical tape

You will also need to buy a 9-volt battery to power your robot.

We'll use the same cues that were introduced in Chapter 7:

Cue	Meaning
✓	**Do** take the action indicated. No particular precision is required.
✕	**Don't** do this! It is a mistake.
✎	**Solder the part** at the point or points indicated.
▭	**Some variation is allowed.** The dimensions are not critical. Stay within 1 or 2 mm of the specified measurement.
◉	**A loose or sloppy fit is OK.** Your interface will work fine if the fit is not tight.

Precautions

Let's review the cautions mentioned in Chapter 7.

✓ Soldering irons get **very hot.**

✕ Do **not** touch exposed metal on a hot soldering iron. Hold it by the insulated handle.

✕ Do **not** lay a hot soldering iron on the work surface or flammable material.

✓ Use a soldering iron holder or lay a piece of wood under but not touching the tip to protect your work surface.

✕ Although the integrated circuits used in the interface are not particularly static sensitive, do **not** handle them by the pins.

✓ In winter and dry weather, touch a large metal object (table, door frame) to discharge any static electricity before handling integrated circuits.

✓ ALWAYS wear safety glasses when performing this work.

Now let's build the interface card and use it to control your robot.

IDENTIFY PARTS

Make sure you have all the parts. Match them against the photograph in Figure 8.3 to be sure they are the correct parts. Refer to Chapter 3 for hints on how to read the values that appear on the resistors.

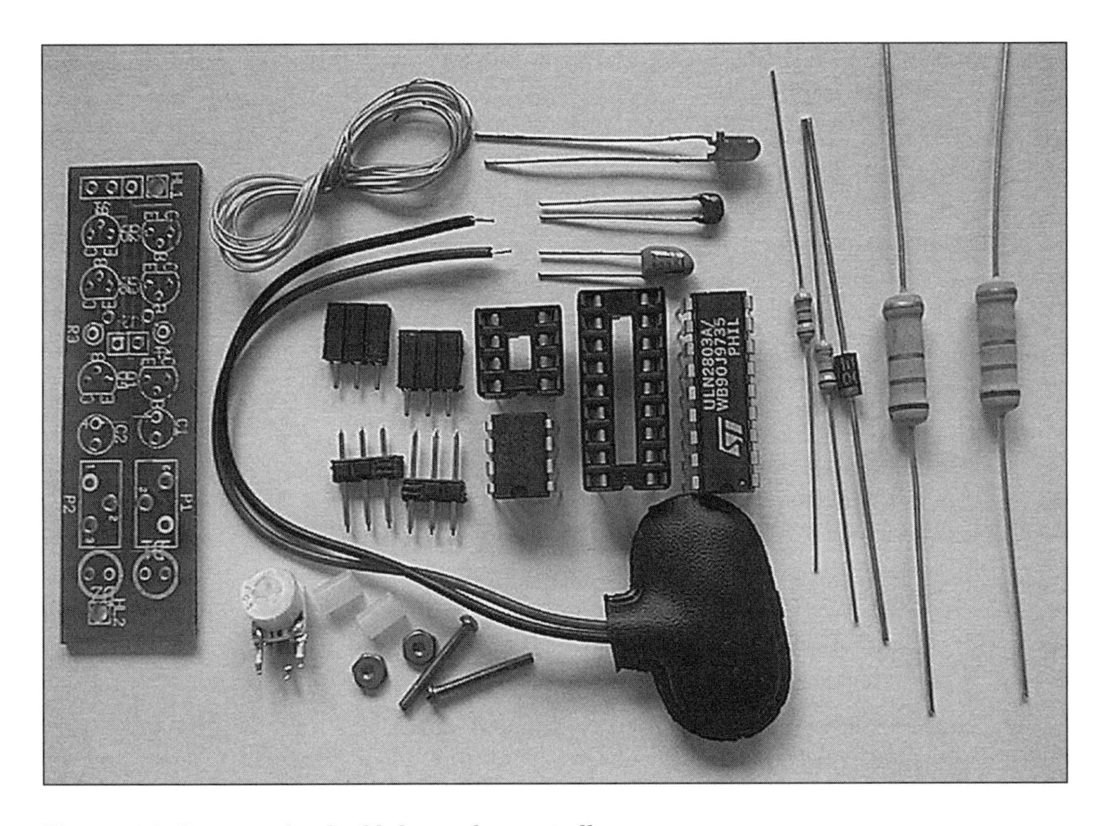

Figure 8.3. Parts used to build the analog controller.

Qty	Name	Where Used or Location on Board
1	Integrated resistor lamp (LED)	board-D1
1	Diode	board-D2
1	555 Timer	board-U1a
1	8-pin DIP socket	board-U1
1	8-transistor Darlington Array chip (ULN2803)	board-U2a
1	18-pin DIP socket	board-U2
2	3-pin socket	board-J1 and battery
2	3-pin jumper post	board-J2 and Stiquito
1	1.0 μf capacitor	board C1
1	0.01 μf capacitor	board C2
1	9V battery connector	battery
2	2 ohm resistors	test components
1	2 Mohm resistor	board R1
1	10 Kohm resistor	board R2
1	1 Kohm resistor	board R3
1	2 Mohm potentiometer	board P1
1	10 cm of 28 AWG wire wrap wire	Stiquito
1	Printed circuit board ST-202	board
2	0-80-¾ inch brass screws	board HL1 & HL2
2	0-80 brass nut	
2	½ inch plastic standoff	

Soldering Skills

Soldering connects components and integrated circuits to the interface card mechanically and electrically. The basic technique for soldering is described in Chapter 7.

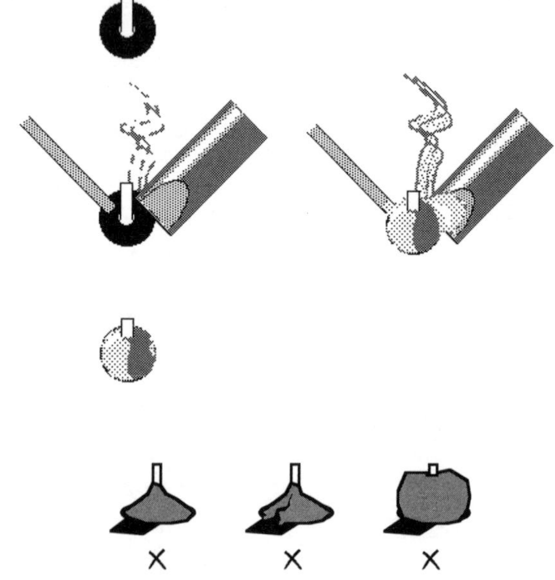

✓ Use the soldering iron's tip to heat the **pad,** not the integrated circuit pin. When the pad is hot, touch the solder to the heated pad, and the solder will flow onto the pad and the pin.

✓ Use just enough solder to wet the pin and cover the pad.

✓ Each solder joint should be bright, shiny, and have flowed evenly around the pin on the pad. The solder on adjacent pads must not touch.

✗ A solder joint should **not** be dull, cracked, or beaded up on the pad.

X A solder joint must **not** cross between two pads, or a pad and a trace. This will create a short circuit. Your interface card will almost certainly **not work correctly.**

After you have finished soldering, you should examine the board for workmanship errors. Check the board for short circuits and broken traces:

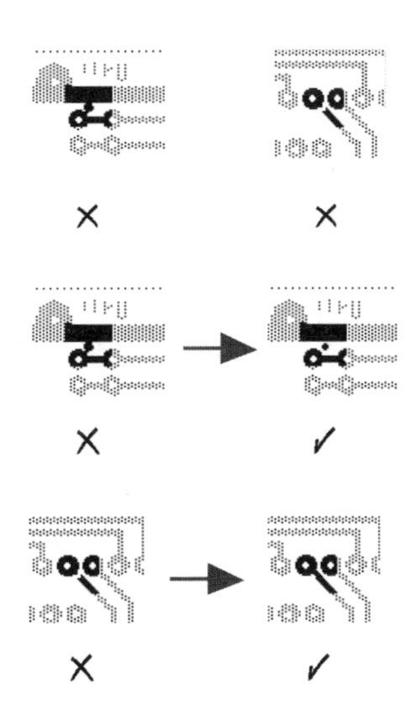

✓ Examine the wiring side of the interface board. Look at places where one trace or pad is near another; check that they do not touch. Look at long traces and near bends; check that the trace is not broken at that point.

✓ If traces or pads touch, but they should not, use the knife to cut the unwanted connection.

✍ If a trace is broken, lightly sand it on either side of the trace, then solder the broken ends together using a piece of fine wire to bridge the gap.

BUILDING THE CIRCUIT BOARD

Printed circuit boards are designed so that, if a component must be inserted by hand, the component's outline and "number" are printed in white on the board. Examine your circuit board. Notice that some letters and numbers are printed near circles and rectangles. These numbers and shapes represent the specific component and orientation of the component. For example, **D1** appears next to a circle. This means that diode 1, integrated resistor lamp (LED), should be inserted in those holes on the board.

Note that some components have a specific orientation and will not work unless you insert them correctly.

- Insert the resistors (R1 through R3) from the component side. Make sure you insert the correct value resistor in the correct location (remember the resistor value discussion in Chapter 3).
- Insert the capacitors (C1 and C2) from the component side, matching the + sign of the board with the + sign on the capacitor.
- Insert the LED (D1) from the component side, matching the flat side of the diode with the flat side on the board drawing (remember the polarity discussion in Chapter 3).
- Insert the diode (D2) from the component side, matching the end of the diode with the bar on the silk screening. See Figure 8.4.

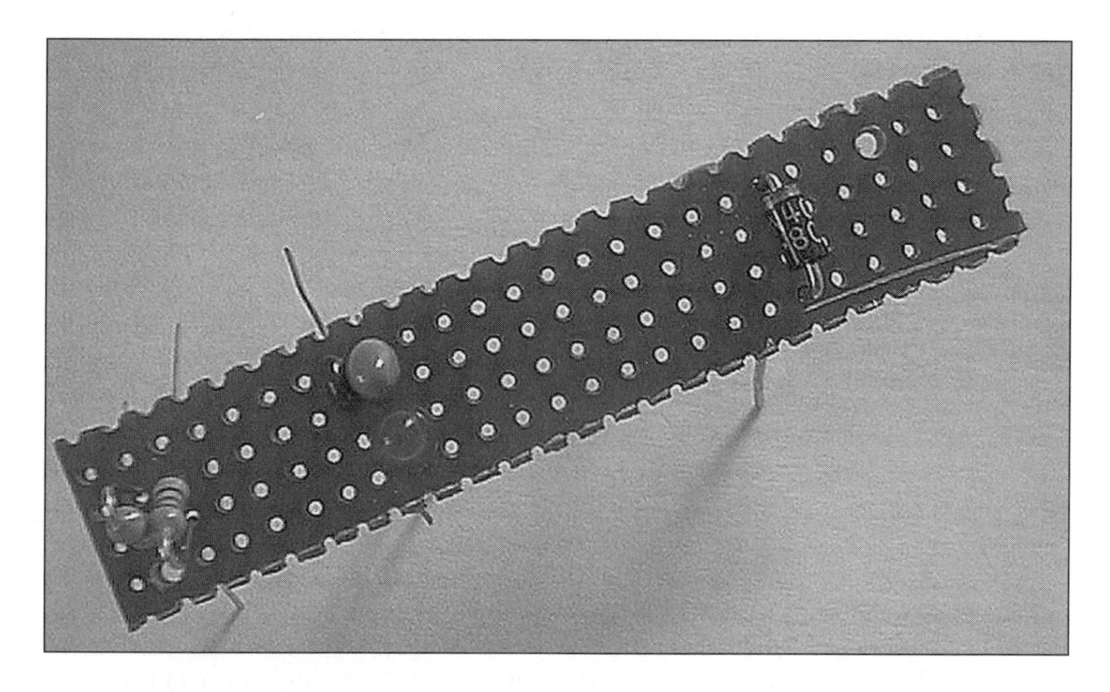

Figure 8.4. Inserting resistors, capacitors, and diodes into the circuit board.

✓ Turn over, bend the component leads (wires) out slightly so that the components do not fall out of the hole.

✍ Solder the leads to the circuit board (Figure 8.5).

✓ Cut the lead wires as close to the solder joints as possible.

✓ Check your work.

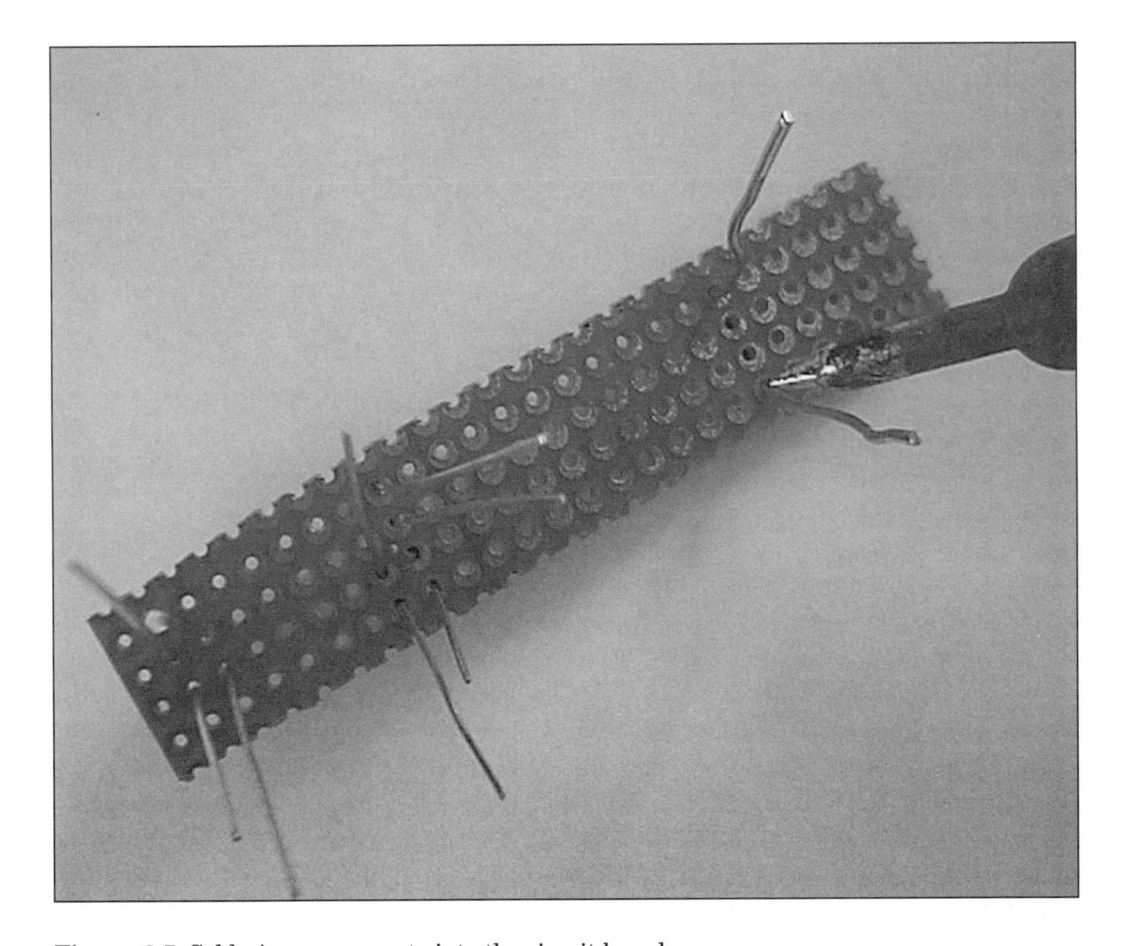

Figure 8.5. Soldering components into the circuit board.

- Insert the 18-pin (U1) and 8-pin sockets (U2) from the component side. Make sure the notch at the top edge of the sockets match the notch in the drawing on the board. See Figure 8.6.
- Insert the socket (J1) and the pin jumper (J2) from the component side.
- Insert the potentiometer (P1) from the component side.
- Turn over, solder, and check your work.

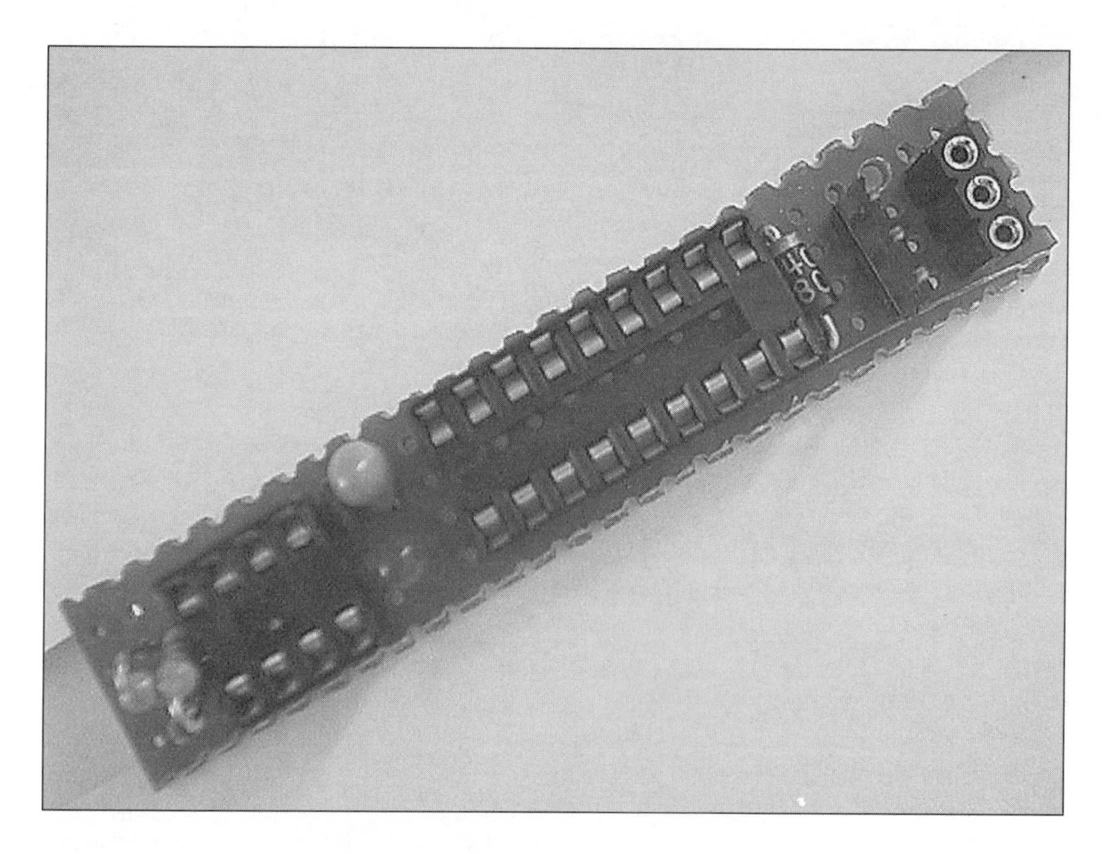

Figure 8.6. Inserting connectors and sockets into the circuit board.

- Solder the 9V battery terminal to the pins of the three-hole socket.
- Cut a small length of wire.
- Solder it to the Stiquito bus bar, then to the center pin of the 3-pin jumper.
- Solder the other two Stiquito tripod control wires to the two outside pins of the 3-pin jumper. See Figure 8.7.

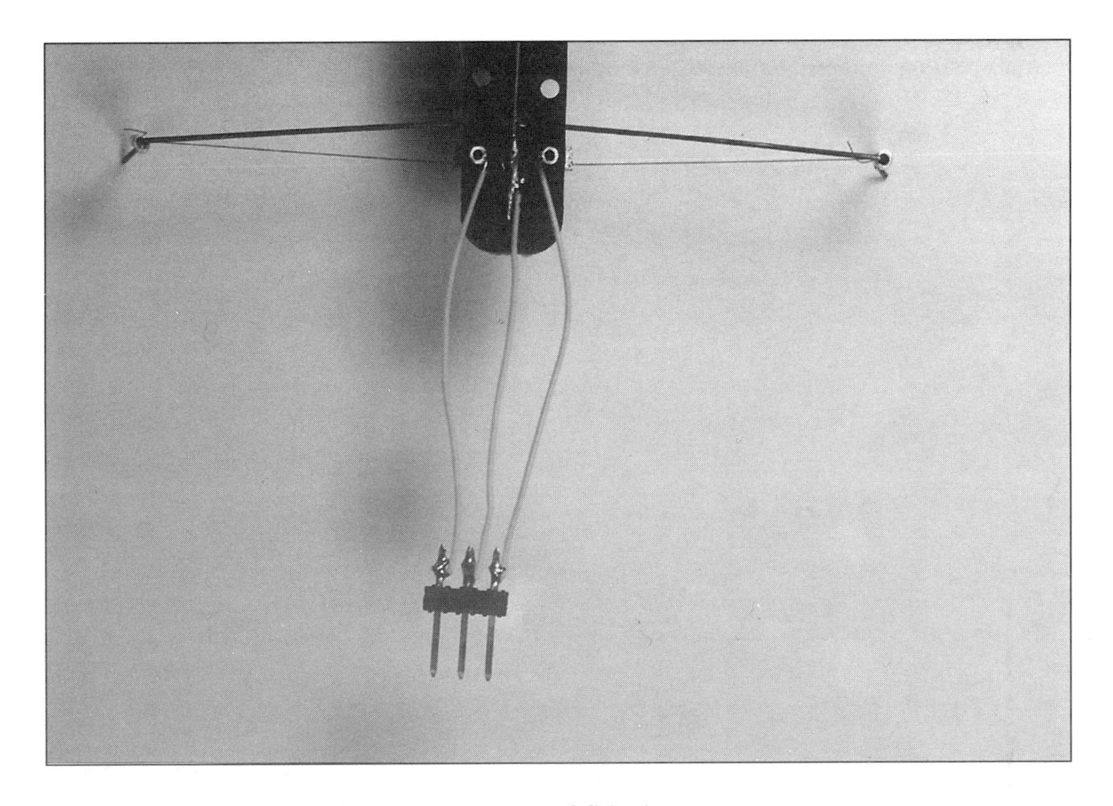

Figure 8.7. Preparing the battery connector and Stiquito connector.

✓ Turn the PCB component side-up.

◉ Carefully insert the ULN3903 chip into the 18-pin socket. Insert the 555 chip into the 8-pin socket. Be careful not to bend the pins. Match the notch on the chip with the notch on the socket.

✓ You may find the chip pins are slightly wider than the socket. One trick to bend all of the pins closer to each other is to lay the chip on its side and gently press down on the other side until the pins bend slightly. This step may take a little practice. See Figure 8.8.

Your analog controller printed circuit board is now complete.

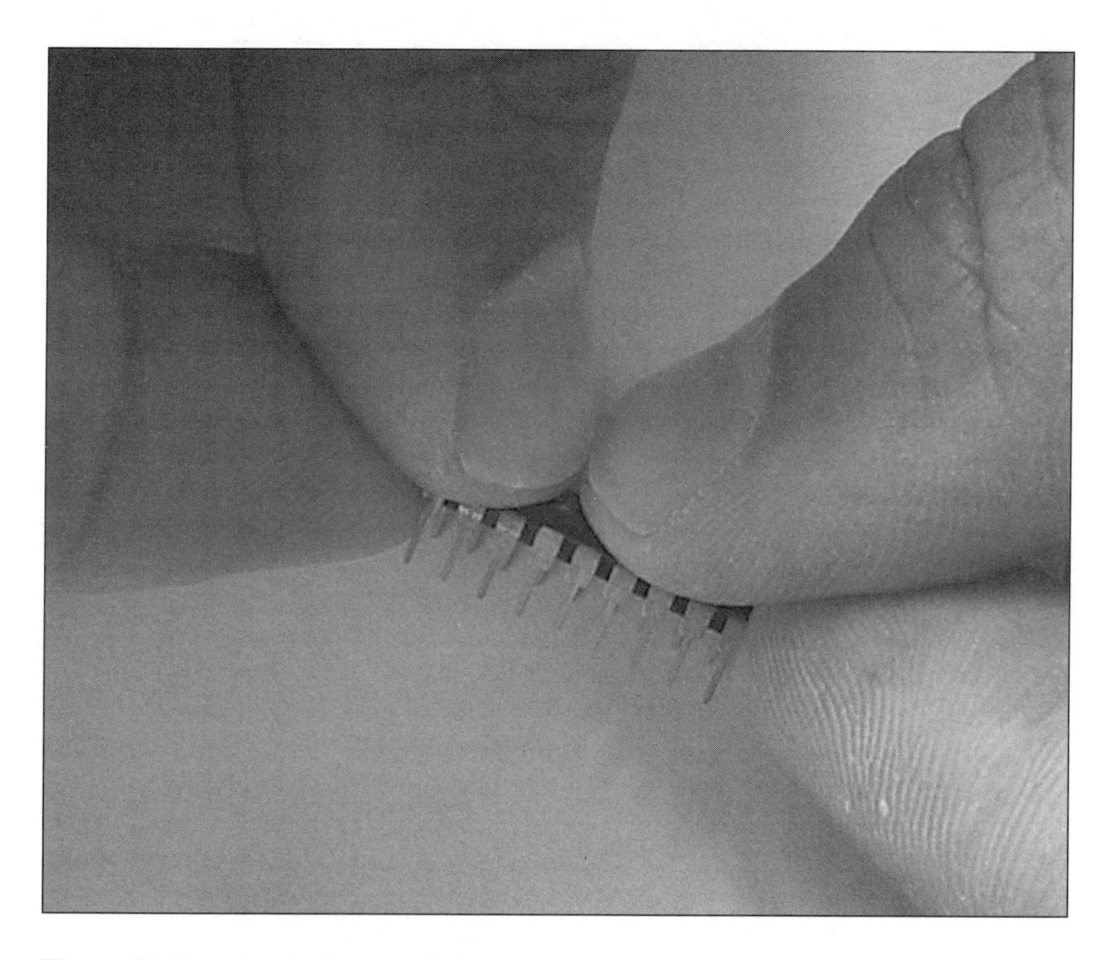

Figure 8.8. Inserting the integrated circuits into the circuit board sockets.

TESTING YOUR CIRCUIT BOARD

Before you hook the circuit board to your robot, test it by simulating the nitinol wires. You will temporarily attach two large resistors instead of the nitinol and then hook the board up to your battery. See Figure 8.9.

✍ Solder one lead of each of two 2-watt resistors together, but make sure the soldered lead can fit into the 3-pin jumper.

◉ Plug the resistor leads into the three holes of the three-hole socket. Make sure the soldered leads are in the middle hole.

✓ Plug the power into the circuit board. Watch carefully for smoke, and ensure the circuit does not emit a burning odor. If it does, remove power immediately and repair your circuit.

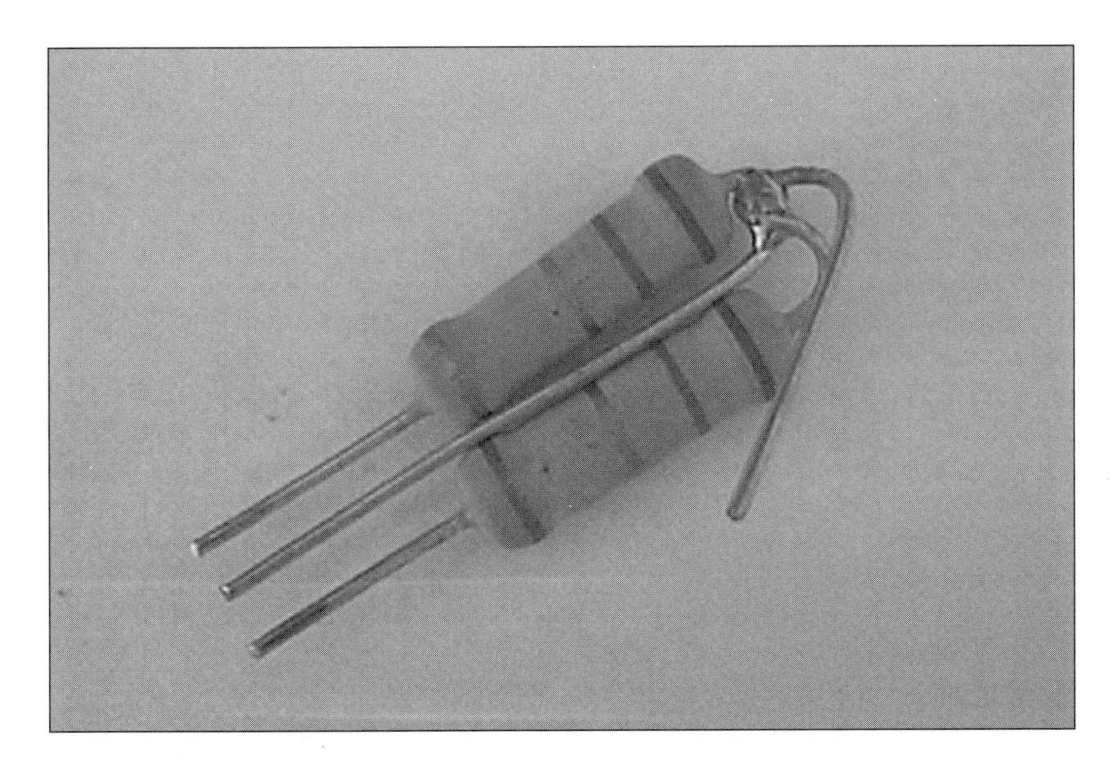

Figure 8.9. Testing the circuit board with a simulated load.

To fine-tune your robot, adjust the potentiometers so that the light blinks for 1½ seconds and rests for 1½ seconds. You will have to do this again when you hook up the robot wires to the circuit.

Now that you are sure your circuit board works correctly, you should plug your robot into the three-hole socket (see Figure 8.10):

✓ Unplug power from the board. Remove the two 2-watt resistors from the board.

✓ Attach the board to the robot, inserting the screws through the board at locations HL1 and HL2. Put the spacers on the screws, then put the screws through the holes in the top of the robot. Attach with the nuts underneath the robot. The three-jumper end of the board goes toward the back of the robot. Make sure there is clearance between the robot and the board (make sure no solder or wires touch the robot).

✓ Plug the power into the board. Plug the robot control socket into the circuit board. Watch carefully for smoke, and ensure the circuit does not emit a burning odor. If it does, remove power immediately.

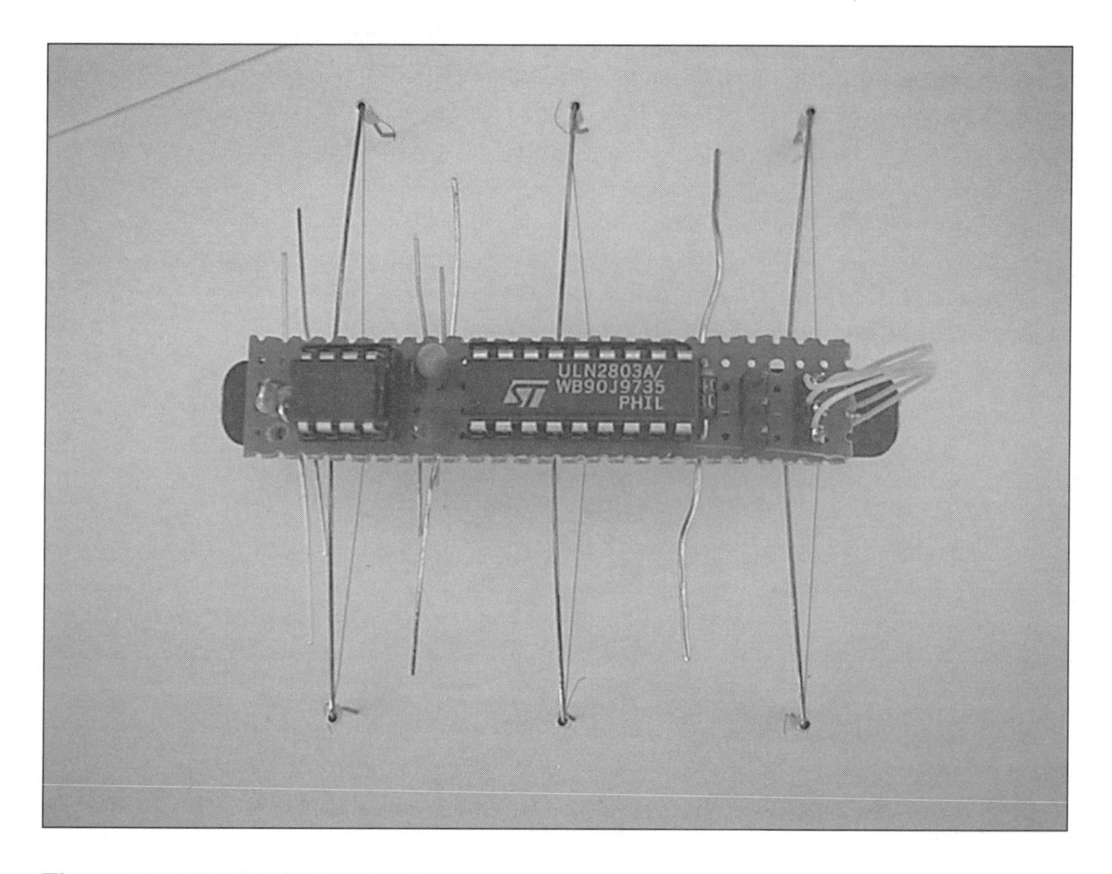

Figure 8.10. The finished product!

To fine-tune your robot, adjust the potentiometer so that the leg actuators contract for 1½ seconds and rest for 1½ seconds.

Once your robot is walking, tape the battery to the top of the circuit with electrical tape.

VARIATIONS OF THE CIRCUIT

You can change the circuit described above to exhibit a light-affected behavior. To make Stiquito walk faster and slower, simply wire a photodiode in line with the 2 Mohm resistor. In earlier versions of this circuit, we have soldered phototransistors on the end of "wire antenna" and used them to follow light (see Figure 8.11). See the first Stiquito book for details on this circuit.[1]

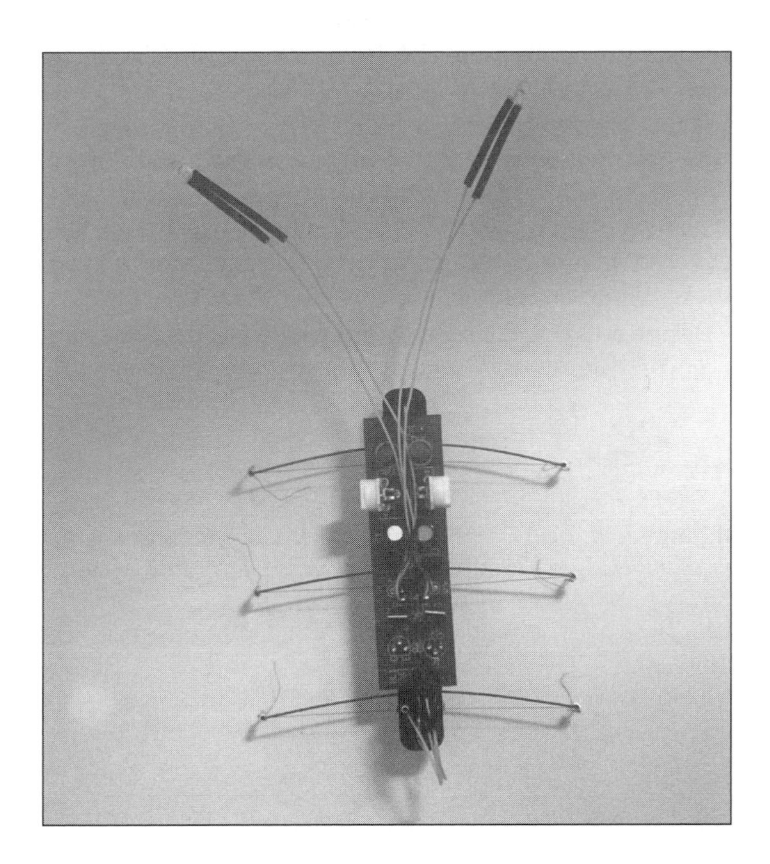

Figure 8.11. An example of a light-finding Stiquito.

One final note is on the use of this driver circuit for walking robots with two ranges of motion. The vertical drivers should not be connected unless additional circuitry is added to disable the horizontal drive circuitry. This is because the oscillator drivers are not synchronized to each other. Without circuitry linking the forward-backward motion of a leg with the raising-lowering of the same leg, you cannot be assured the leg will be touching the ground when it is moved forward.

EXERCISES

1. In the circuit shown in Figure 8.1, the *duty cycle,* or the time the voltage of the output is high versus low, is about 50% (1.733/1.739). What resistance would R2 need to be to make a 66% duty cycle?
2. To test the circuit board, we used a 2.0 ohm resistor to simulate the nitinol wires. Based on this and the formula $V = IR$ introduced in Chapter 3, what are the expected resistance and current of each nitinol wire.
3. This circuit has been built with a 6 volt J-type battery. Under this condition, what are the expected resistance and current of each nitinol wire.
4. The printed circuit board provided in the kit is a two-layer board. Examine the board and identify the *traces,* or electrical paths on the board. Identify the traces that go between two pins of the 555 timer with the resistor and potentiometer in-line. Where would you put the phototransistor in this circuit?
5. (Difficult) Design additional circuitry that would be added to two of these circuits boards to control Stiquito walking with two degrees of freedom.

BIBLIOGRAPHY

1. Conrad, J. M., and J. W. Mills. 1997. *Stiquito: Advanced experiments with a simple and inexpensive robot.* Los Alamitos, Calif.: IEEE Computer Society Press.

Chapter 9

The Future of Stiquito and Walking Robots

Jonathan W. Mills and James M. Conrad

We admit to being pleased that Stiquito is available in two books that collect material from many authors. After having lived with this little robot for many years, we would like to add a few points and repeat a few ways that you can use Stiquito successfully.

BUILDING STIQUITO

The book advises you to read the instructions first. This is essential. But we will add that it greatly helps to have, or to practice, exercises that improve your manual dexterity. Mark Tilden, famous roboticist, suggested practicing these skills by bending a paper clip as straight as a pin. If it rolled on an incline, then you could claim a high ~Q. (That's "Tilde-Q." Like IQ. Get it?) Surprisingly, this helps. You get a sense of how your fingers can fine-twiddle (or fine-tease) a stubborn piece of metal into a desired shape.

If you buy a book on nitinol from Mondo-Tronics or Dynalloy, each will emphasize—as do our books—that nitinol absolutely must be taut, without any slack, and with an unmovable anchored attachment to function. Over the past few years loose nitinol has been the major reason above all others that "the darned robot did not work." On the other hand, a working Stiquito is a thing of beauty. Dr. Mills still recalls seeing Professor Russ Fish's (University of Utah) Stiquito with legs with two degrees of freedom lifting each leg with a determined and almost fussy precision as it walked in a tripod gait.

You can use the kit body to build a version of the robot with two degrees of freedom. You will need to buy extra nitinol, though, and also need to build a controller to match.

OPERATING STIQUITO

First of all, forget the movie "Runaway," with its six-legged robot insects running around at light speed. Forget speed. Forget carrying sensors that you found in a scrapped laser printer. Stiquito is slow. It can't carry much weight. It doesn't have much space to attach the gadgets you know and love (ultrasonic ranging gear, video cameras, etc.). If you need or want this sort of thing, don't even think of using Stiquito. But, you can add sensors to do useful experiments with the robot:

- Photodiodes
- Simple bump sensors ("whiskers")
- Dead reckoning sensors (leg sweep)
- "Virtual reality" sensors (with a PIC or BASIC stamp microcontrollers)

All of these are lightweight and involve simple mechanical actuation, or none at all. Putting a BASIC Stamp microcontroller on board, with a thin wire power tether, is the best compromise we have found so far between independent control and power consumption. The Stamp also lets you create a virtual environment for the robot. The trick is to use another Stamp to send bit-serial signals to whiskers in the environment. The robot "reads" these signals with a wire that is an open-collector serial input. When two wires touch, even for a few milliseconds, a message of several bytes—six to 20—can be transmitted reliably. Voila! A virtual reality for Stiquito, useful for experiments in foraging or cooperative behavior, for example.

A second type of virtual reality sensor is a video camera focused on the robot arena. Two alumni of Dr. Mills' lab, Jason Almeter and Perry Wagle, developed video capture and image recognition routines that would let one camera act as a sensor for multiple robots. The control loop would have been closed by sending information about the environment back to each robot via a serial tether, filtering the information to limit the robot's sensed environment. Unfortunately, the colony project remains to be completed.

Of course, simple controllers are useful. One of the simplest is to connect the robot directly to the printer port of an IBM PC, as shown in Chapter 7. This method was first suggested by Dennis Clarke of Oak Ridge National Laboratory, who sent us the "harness" used to drive Stiquito. Although this may be possible for your PC, we suggest using the PC output lines for driving transistors and a non-PC supply for the increased current to drive the legs.

The first Stiquito book[1] contains several chapters devoted to controlling the robot. Included as ideas are:

- A PC-based controller to drive up to 12 nitinol wires. This is useful if you build a Stiquito with two degrees of freedom and want to control each nitinol wire separately.
- A microcontroller-based controller. This circuit gives autonomous control of the nitinol wire without being attached to a PC.
- An infrared communications controller. This circuit allows one to send infrared commands from a handheld remote to control the walking gait of Stiquito.
- A field programmable gate array (FPGA) controller. A circuit is designed and burned into a FPGA. This does not allow the controller to be reprogrammed but reduces the weight of the circuit carried on Stiquito's back.

There have been suggestions about using temperature sensors to control a pulse width modulated driver or using the change in resistance. They won't work unless you are extremely careful—and lucky. Nitinol wire is so thin that most sensors are too bulky to measure the rapid changes in temperature. Similarly, changes in resistance are difficult to measure to obtain real-time control (we have been there and tried that in the lab). What does work is to calibrate a PWM driver for your actuator, using the data from Dynalloy (for example) to actuate the nitinol, then reduce the pulse width (or frequency) to hold the wire in its contracted state. We have done this using a Stamp microcontroller from Parallax, Inc., but when the wire begins to resonate audibly—to "sing"—it may break.

OTHER STIQUITO-LIKE ROBOTS

Stiquito II

Stiquito II is an improved Stiquito. It is larger and modular, with an articulated body and legs with two degrees of freedom, and it carries more weight. It is easier to handle than Stiquito because it is larger. Its articulated body allows it to walk up and down shallow stairs, climb small inclines, negotiate rough terrain, and turn to avoid obstacles. A photograph of Stiquito II is in Chapter 1.

Stiquito '97

The Stiquito '97 robot, shown in Figure 9.1, is built on a printed circuit board, uses an on-board BASIC Stamp 2 microcontroller for control, and is tethered only to power. The design was intended to be implemented using a printed circuit board as its body, which reduces assembly time to about two hours. The music wire legs and nitinol wires are attached directly to the printed circuit board, as shown in Figure 9.2. A small program can be downloaded to the BASIC Stamp 2 to control the gait and to enable communications with other Stiquito '97 robots.

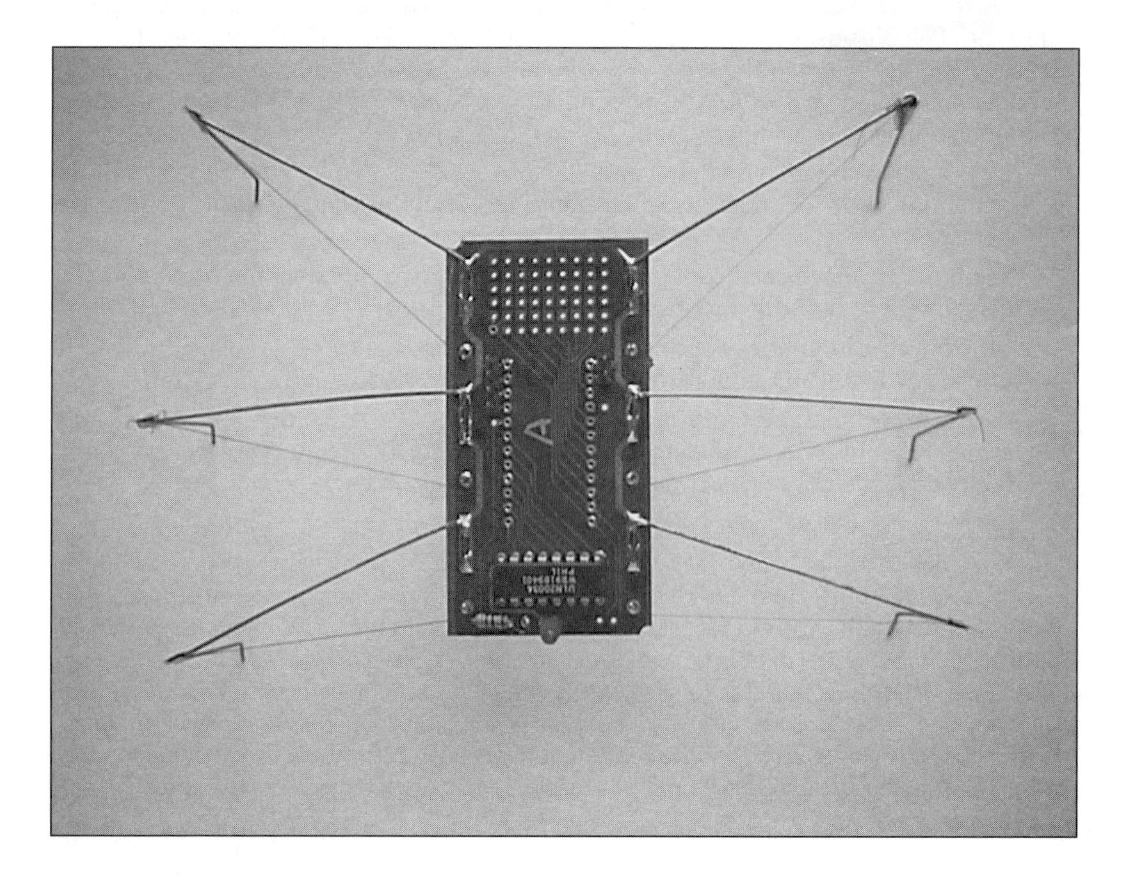

Figure 9.1. Stiquito '97.

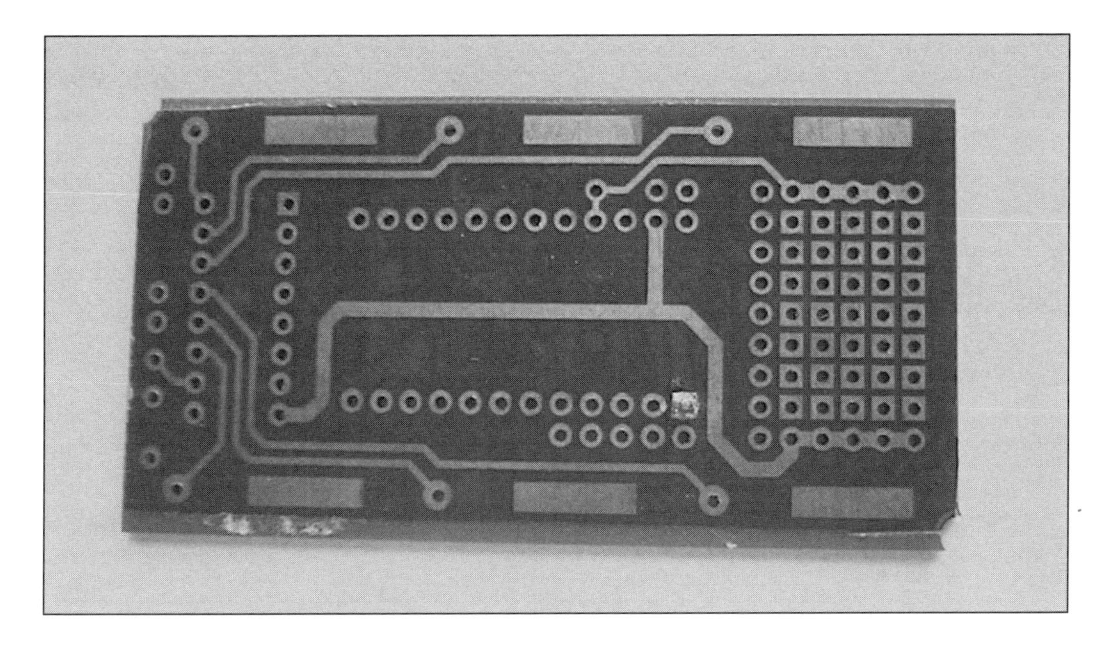

Figure 9.2. The Stiquito '97 printed circuit board.

Boris

Boris is another walking robot similar to Stiquito. Boris uses eight separate nitinol wires to create its motion: one to make the forward and backward motion on each of its legs (six total), and two wires for lifting alternating legs with each step. It is made from balsa wood, plastic, music wire, and hardware fasteners like screws and nuts. Boris is connected to a wired harness and controlled by a PC parallel port card, similar to the card you built in Chapter 7. A photograph of Boris is provided in Chapter 4.

Other Nitinol-Based Creatures

Nitinol-propelled robots need not be restricted to six-legged, bug-like designs. Roboticist Mark Tilden has designed hundreds of robots, many of which do not look like bugs. One such creature, Nitecrawler 1.0, shown in Figure 9.3, uses nitinol is such a way that the small contraction of the wire causes a large movement. The robot resembles a snail in appearance and movement. Most impressive is that Nitecrawler moves 3 cm with each "step," yet uses only 15 cm of nitinol wire.

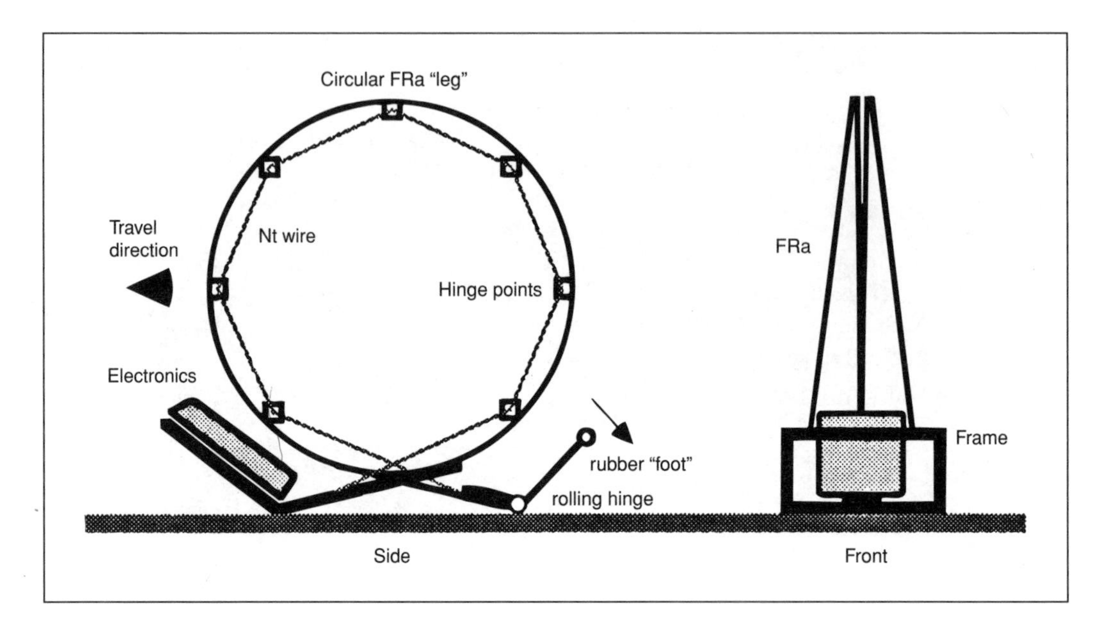

Figure 9.3. A Design Schematic of Nitecrawler 1.0.

THINGS TO DO WITH STIQUITO

Stiquito has been used by some people repeatedly for classes over the past several years. Prof. Vijay Kumar at the University of Pennsylvania uses it to teach a design class in mechanical engineering. The robot provides an inexpensive way to illustrate the interface between digital control and a mechanical plant (i.e., the controlled analog system).

Dozens of informal science education groups have built Stiquitos, as have many high school and undergraduate students doing science projects. The typical project involves studying insect gaits. One example of a project is posted on the IEEE Computer Society Stiquito Supplemental page.[2] Max Eskin, a high school student, built a Stiquito with control wires to each leg. He then experimented with different gaits, including moving one, two, and three legs at a time.

This is definitely a field with potential for further study! One of Jonathan's graduate students, Gary Parker, has invented the idea of Cyclic Genetic Algorithms (CGAs) that deal with the evolution of repetitive processes. He has published numerous papers and explored the topic in his doctoral dissertation. We have heard of Stiquito Olympics that include events for distance, straightness, and other competitions. We were also amused by some ideas suggested by a local researcher in semiotics, who thought that Stiquito ballet, archeology ("excavate" the junk heap left from building the robot and try and deduce what it looked like), and other such fantastic events could be fun. Ryan Keating, son of one of Dr. Mills' colleagues, suggests studying the gaits of water striders. Certainly this would be a challenge: first the robot would have to be buoyed up, then the foot would have to be constructed. One might want to use heavy oil instead of water. John Nagle has suggested robot fish; Stiquito would not be a good candidate but might lead to an idea for a swimmer with a nitinol actuated spine.

THE FUTURE OF STIQUITO

What can Stiquito be used for in the future? As suggested in several futuristic movies, legged robots may have value for certain applications. Surely Stiquito could have some applications, even though it is small.

- Mark Tilden has suggested biodegradable insect robots for pest control on farms.[3] Instead of spraying environmentally hazardous chemicals on crops, one could release thousands of small Stiquitos in the fields. The robots would seek out and "eat" harmful pests. We suggest relying on sunlight for power. At the end of the growing season, the robots would be tilled into the earth and decompose, thus supplying nutrients for the soil.
- Instead of sending one, very expensive wheeled vehicle to Mars, why not send thousands of small Stiquitos? Each robot would have sensors and a radio to transmit data to the "ant hill," or the home base on Mars. These would also be solar powered and could walk and collect data for years. How far could they walk? You can figure that out yourself in an exercise at the end of this chapter.
- In schools of the future, scavenger Stiquitos could leave their home base in the cafeteria and wander the floor in search of crumbs and other bugs to eat! In fact, why limit them to schools? They could even be used in your kitchen!
- Stiquitos could be inserted into one end of a large, clogged pipe to dig their way through the clog to the other side.
- More nimble variations of Stiquito, equipped with microphones and cameras, could be sent into the rubble of a building to look for survivors.
- Grass-chomping Stiquitos could cut the lawn while you sleep.

The future of Stiquito is only limited by your imagination!

IN CLOSING

A colleague of Jonathan Mills observed that the longevity of Stiquito is probably due to two factors.

- It is simple to build.
- Everybody believes s/he can build a better one!

We agree. As an inexpensive introduction to robotics, this little 'bot can get you started. But you don't have to stop. Dave Braun, another graduate of Indiana University who is now working for Caterpillar, Inc., designed the six-legged Servobot that is sold by Lynx Motion, Inc. He wanted a faster, bigger robot that could carry more weight—and built it!

In nature Stiquitos would be greedy, blind, crippled, dumb ants. They consume electricity voraciously. They have no sensors or effectors (at least as described in this book). Their legs lack the numerous degrees of freedom of those of an ant. Their controllers, whether a PC, an analog circuit, or a microcontroller, are overwhelmingly large and clunky as compared to the ganglia that serve an ant (at least to accomplish an ant's tasks).

So why build Stiquitos?

We build them because they are simple, they are fun, and they let us investigate robotics without getting a million-dollar grant. We build Stiquito because we have yet to construct anything the size of an ant that has the capability of an ant. But we are working on it.

We encourage all readers of this book to improve on Stiquito. If you build a better, inexpensive ant-like robot, we will stand in line to get one!

EXERCISES

1. List all the types of inputs you could want for an electronic robot. For example, you may want the ability to know when you touch something.
2. Electronics, and especially nitinol, require electricity to run. List all the different sources of electricity you could provide for a robot.
3. Describe changes you would need to make to Stiquito to allow it to walk with two degrees of freedom.
4. Describe the changes you would need to make to Stiquito electronics to control it to walk with two degrees of freedom. Assume you are using the PC parallel port controller described in Chapter 7.
5. Describe the gait you would need to control a Stiquito with two degrees of freedom.
6. A Stiquito robot walks unimpeded in a straight line on Mars. If it walks at a rate of 10 cm per minute for 12 hours and 0 cm per minute for 12 hours of each earth-day, how far will it walk in one earth year? Express your answer in kilometers.
7. Repeat Exercise 6, but instead of one earth year, use one Martian year. You will need to search astronomical sources for the definition (in earth days) of one Martian year.
8. (Difficult) How can you improve Stiquito? What new design is needed to make it work better and still be inexpensive?

BIBLIOGRAPHY

1. Conrad, J. M., and J. W. Mills. 1997. *Stiquito: Advanced experiments with a simple and inexpensive robot.* Los Alamitos, Calif.: IEEE Computer Society Press.
2. IEEE Computer Society. 1999. Stiquito Supplemental Web Page: http://computer.org/books/stiquito
3. Conrad and Mills, 1997.

Appendix A

Author Biographies

James M. Conrad

James M. Conrad received his bachelor's degree in computer science from the University of Illinois, Urbana, and his master's and doctorate degrees in computer engineering from North Carolina State University.

He is currently an engineer at Ericsson, Inc., and an adjunct professor at North Carolina State University. He has served as an assistant professor at the University of Arkansas and as an instructor at North Carolina State University. He has also worked at IBM in Research Triangle Park, North Carolina, and Houston, Texas; at Seer Technologies in Cary, North Carolina; at MCI in Research Triangle Park, North Carolina; and at BPM Technology in Greenville, South Carolina.

Dr. Conrad is a member of the Association for Computing Machinery, Eta Kappa Nu, and IEEE Computer Society. He is also a Senior Member of IEEE. He is the author of numerous articles in the areas of robotics, parallel processing, artificial intelligence, and engineering education. Dr. Conrad can be reached at the following address:

Ericsson, Inc.
7001 Development Drive
Research Triangle Park, NC 27709
Phone: 919-472-6178
Fax: 919-472-6515
E-mail: jconrad@stiquito.com (preferred contact method)

Jonathan W. Mills

Jonathan W. Mills received his doctorate in 1988 from Arizona State University. He is currently an associate professor in the Computer Science Department at Indiana University and director of Indiana University's Analog VLSI and Robotics Laboratory, which he founded in 1992. Dr. Mills invented Stiquito in 1992 as a simple and inexpensive walking robot to use in multirobot colonies and with which to study analog VLSI implementations of biological systems. In 1994 he developed the larger Stiquito II robot, which is used in an eight-robot colony in his laboratory. Since 1992 Indiana University has distributed more than 3,000 Stiquito robots, leading to the idea for this book.

Dr. Mills is currently researching biological computation in the brain using tissue-level models of neural structures implemented with analog VLSI field computers. Field computers offer a powerful but simple paradigm for adaptive robotic control. They are small and light enough to be carried by Stiquito, yet still perform sensor fusion and behavioral control.

Dr. Mills has written a series of papers on his analog VLSI and robot designs; he has one patent with several others pending and applied for on his work. He also freely admits that Stiquito is just the start of what he hopes will be a series of improved and functional miniature robots, and he encourages the readers of this book to be inspired to design and build them. Dr. Mills can be reached at the following address:

215 Lindley Hall, Computer Science Department
Indiana University
Bloomington, IN 47405
Phone: 812-855-6486
Fax: 812-855-4829
E-mail: stiquito@cs.indiana.edu

Allan R. Baker

Allan R. Baker received his bachelor's and master's degrees in electrical engineering from the University of Arkansas. His master's thesis applied artificial neural networks to the problem of reducing disturbances in control systems. While working on his undergraduate degree, he did four semesters of cooperative-education work with NASA's Marshall Space Flight Center in Huntsville, Alabama. His experiences included electrical parts' failure analysis and assisting in the study and fabrication of low-cost solar cells. The university awarded him the Porter Stone Outstanding Co-op Award in 1993. For his senior design project he designed and built a mobile robot that used infrared sensors and a Motorola HC11 microcontroller to navigate its surroundings. Since completing his master's degree in 1996 he has worked as a systems engineer for Texas Instruments Defense Systems, which Raytheon Systems Company acquired in 1997. His duties have included the definition of requirements and performance analysis for the design of a multiprocessor embedded computer system as well as control and signal processing algorithm design and analysis. His interests include neural networks, computer architecture, machine intelligence, signal processing, control systems, and robotics. Mr. Baker can be reached at:

E-mail: arbaker@gte.net

Wayne Brown

Wayne Brown is president and founder of Dynalloy, Inc., the manufacturer of Flexinol™ actuator wire. He has spent considerable time developing ways, both mechanically and electrically, to use these wires correctly. He has developed numerous products of his own and assisted in the design of several hundred others. His own inventions include robotic carts, simple CNC machines, toy doll movements, and mechanical butterflies. Mr. Brown has been in the computer and robotics field for more than 27 years. He can be reached at:

3194-A Airport Loop Drive
Costa Mesa, CA 92626-3405
Phone: 714 436-1206
Fax: 714 436-0511
E-mail: flexinol@dynalloy.com
Web address: http://www.dynalloy.com

James G. Martin

James G. Martin taught himself to draw at a very early age. His love for cartooning prompted him to seek formal training and a bachelors degree in Illustration at Northern Illinois University. He has designed holiday gift ware and art for education toys and has worked with many licensed properties such as Disney, The Muppets, and Looney Tunes. He is currently Art Director for a leading agency that designs and produces Happy Meal Toys for McDonald's. James lives happily in Lombard, Illinois, with his wife and son. Mr. Martin can be reached at:

E-mail: jmartin@stiquito.com.

Jon E. Pedersen

Jon E. Pedersen earned his Ph.D. in 1990 from the University of Nebraska at Lincoln after a successful career as a chemistry and physics teacher in eastern Nebraska. He has published more than 40 articles and book chapters. His academic work has taken him to numerous foreign countries including extended work in Bolivia, South America. Dr. Pedersen is currently the Associate Dean for Research, Graduate Studies and Professional Development in the School of Education, East Carolina University. Prior to his appointment to this position, he was a faculty member in the Department of Science Education at East Carolina University. He also served as an associate professor at the University of Arkansas. Dr. Pedersen may be reached at:

Associate Dean, School of Education
Speight 154
Greenville, NC 27858-4353
Phone: 252-328-4260
Fax: 252-328-4219
E-mail: pedersenj@mail.ecu.edu

Appendix B

Sources of Materials for Stiquito

This appendix lists some suppliers of Stiquito parts, robotics kits, and electronics. You may find many Stiquito supplies and tools at your local hobby supply store. Check local electronics supply stores for other supplies. This section is not an endorsement of these companies, but is provided to make your supplier search easier.

All Electronics Corporation
P.O. Box 567
Van Nuys, CA 91408
Phone: 888-826-5432, 818-904-0524; Fax: 818-781-2653
E-mail: allcorp@allcorp.com; Web address: http://www.alleletronics.com
Surplus dealer of boards, components, and assemblies.

Digi-Key
701 Brooks Ave. South, P.O. Box 677
Thief River Falls, MN 56701-0677
Phone: 800-344-4539 or 218-681-6674; Fax: 218-681-3380
Web address: http://www.digikey.com

Dynalloy, Inc.
3194-A Airport Loop Drive
Costa Mesa, CA 92626-3405
Phone: 714-436-1206; Fax: 714-436-0511
E-mail: flexinol@dynalloy.com; Web address: http://www.dynalloy.com
Supplier of nitinol wire, trade named Flexinol. Flexinol can be ordered in lengths of 1 meter or more.

Edmund Scientific
101 E. Gloucester Pike
Barrington, NJ 08007-1380
Phone: 800-728-6999 or 609-547-3488; Fax: 856-547-3292
Web address: http://www.edsci.com
Sells optical components, science kits, surplus motors, and robot kits.

Hamilton Hallmark/Avnet
Phone: 800-332-8638; Fax: 800-257-0568
Web address: http://www.hh.avnet.com
Distributor for many semiconductor manufacturers.
Check the web site for your state or country contact.

IEEE Computer Society Press
10662 Los Vaqueros Circle, P.O. Box 3014
Los Alamitos, CA 90720-1314
Phone: 800-CS-BOOKS (800-272-6657) or 714-821-8380; Fax: 714-821-4641
E-mail: cs.books@computer.org; Web address: http://www.computer.org/cspress
Source for Stiquito books and kits.

IS Robotics and Artificial Creatures (a subsidiary of IS Robotics)
22 McGrath Hwy., Ste. 6
Somerville, MA 02143
Phone: 617-629-0055; Fax: 617-629-0126
E-mail: info@isr.com; Web address: http://www.isr.com
Source for robots and sensor systems for education and research.

Jameco Electronic Components
1355 Shoreway Rd.
Belmont, CA 94002-4100
Phone: 800-831-4242 or 650-592-8097; Fax: 650-592-2503
E-mail: sales@jameco.com or international@jameco.com
Web address: http://www.jameco.com
Supplier of electronic components.

K&S Engineering
6917 W. 59th St.
Chicago, IL 60638
Phone: 773-586-8503; Fax: 773-586-8556
Supplier of music wire and aluminum, copper, and brass tubing. Minimum order of $80.00. (Hobby shops also carry these items.)

LEGO DACTA
Lego Educational Department
P.O. Box 1600
Enfield, CT 06083
Phone for education purchases: 860-749-2291; Fax 860-763-0522
Phone for home purchases: 800-453-4652; Fax 888-329-5346
Web address:http://www.lego.com/dacta
Sells components needed for quickly building robot prototypes; educational department sells primarily to schools. Also check out Lego Mindstorms at http://www.legomindstorms.com

Micro Fasteners
110 Hillcrest Rd.
Flemington, NJ 08822
Phone: 800-892-6917 or 908-806-4050; Fax: 908-788-2607
E-mail: microf@blast.net; Web address: http://www.microfasteners.com
Supplier of brass #0 screws, nuts, and washers used on the Stiquito body.

Micro-Robotic Supply, Inc.
101 Pendren Place
Cary, NC 27513-2225
E-mail: sales@stiquito.com; Web address: http://www.stiquito.com
Supplier of the Stiquito Repair Kit, the Stiquito Screws Kit, and other Stiquito accessories.

Mondo-Tronics
4286 Redwood Hwy., #226
San Rafael, CA 94903
Phone: 800-374-5764 or 415-491-4600; Fax: 415-491-4696
E-mail: info@mondo.com; Web address: http://www.robotstore.com
Supplier of nitinol and Flexinol™ wire, robots, robotic books, videotapes, even robotic artwork.

Mouser Electronics
2401 Hwy. 287 N.
Mansfield, TX 76063-4827
Phone: 800-34-MOUSER (800-346-6873) or 817-483-6888; Fax: 817-483-6899
E-mail: sales@mouser.com; Web address: http://www.mouser.com
Wide selection of electronic components. Regional distribution centers; will fax detailed specifications. Accepts small orders.

New Micros, Inc.
1601 Chalk Hill Rd.
Dallas, TX 75212
Phone: 214-339-2204; Fax: 214-339-1585
Single-board computer uses MC68HC11 chip; Forth language in ROM.

Newark Electronics/Farnell
U.S. and Canada catalog requests: 800-4-NEWARK (800-463-9275)
Central and South America: 915-772-6367
Pacific Rim: 619-691-0141
Europe, MiddleEast, Africa: 0044 113 279 3100
Web address: http://www.newark.com
Check telephone directory for a local sales office.
Distributor of electronic components.

Parallax, Inc.
3805 Atherton Road, Suite 102
Rocklin, CA 95765
Phone: 888-512-1024 or 916-624-8333; Fax: 916-624-8003
http://www.parallax.com, http://stampsinclass.com
Supplier of the BASIC Stamp microcontroller, educational curriculum, and robotics kits. Supplier of Stiquito kits described in this book.

Parts Express
725 Pleasant Valley Drive
Springsboro, OH 45066-1158
Phone: 800-338-0531 or 415-743-3000; Fax: 415-743-1677
E-mail: sales@partsexpress.com; Web address: http://www.partsexpress.com
Supplier of electronics, tools, hardware, and supplies.

Plastruct, Inc.
1020 S. Wallace Pl.
City of Industry, CA 91748
Phone: 626-912-7017 or 800-666-7015; Fax: 626-965-2036
E-mail: Plastruct@plastruct.com; Web address: http://www.plastruct.com
Supplier of plastic stock used in Stiquito II, Boris, and SCORPIO. Plastruct offers a 30 percent educational discount to instructors and schools using a purchase order.

Radio Shack
Phone: 800-THE-SHACK
Web address: http://www.tandy.com
National chain; consult telephone directory for nearest dealer. Offers a variety of electronic components from local distributors. For mail order, see Tandy Electronics.

RadioShack.com
PO Box 1981
Fort Worth, TX 76101-1981
Phone: 800-877-0072; Fax: 800-813-0087
Web address: http://www.Radioshack.com
A division of Radio Shack; distributes (by mail order) a wider variety of parts than are available in Radio Shack stores.

Small Parts, Inc.
13980 NW 58th Court, P.O. Box 4650
Miami Lakes, FL 33014-0650
Phone: 800-220-4242 or 305-558-1255; Fax: 800-423-9009 or 305-558-0509
Web address: http://www.smallparts.com
Supplier of metal, plastics, tools, and hardware.

Solarbotics
179 Harvest Glen Way NE
Calgary, AB
Canada T3K 4J4
Phone: 403-818-3374; Fax: 403-226-3741
E-mail: sales@solarbotics.com; Web address: http://www.solarbotics.com
A source for solar-powered robot kits.

Appendix C

Using Screws Instead of Crimps in Stiquito

The Stiquito robot body was designed so that you could also assemble it using screws instead of aluminum crimps. If you wish to use screws instead of crimps, use the sets of holes on the body that are offset slightly. The offset holes work such that you can wrap the nitinol in the same direction as the screw thread. The nitinol is then anchored at the same distance from the legs on each side of the body. I use one brass screw ($^{5}/_{16}''$ × #0-80), one brass washer (#0), and two brass nuts (#0-80) for each hole. The round screw head faces down, and one of the nuts on top tightens the screw. The other nut anchors the control wire to the screw. This assembly is illustrated in the figure below. See the list of suppliers in Appendix B for sets of screws, washers, and nuts.

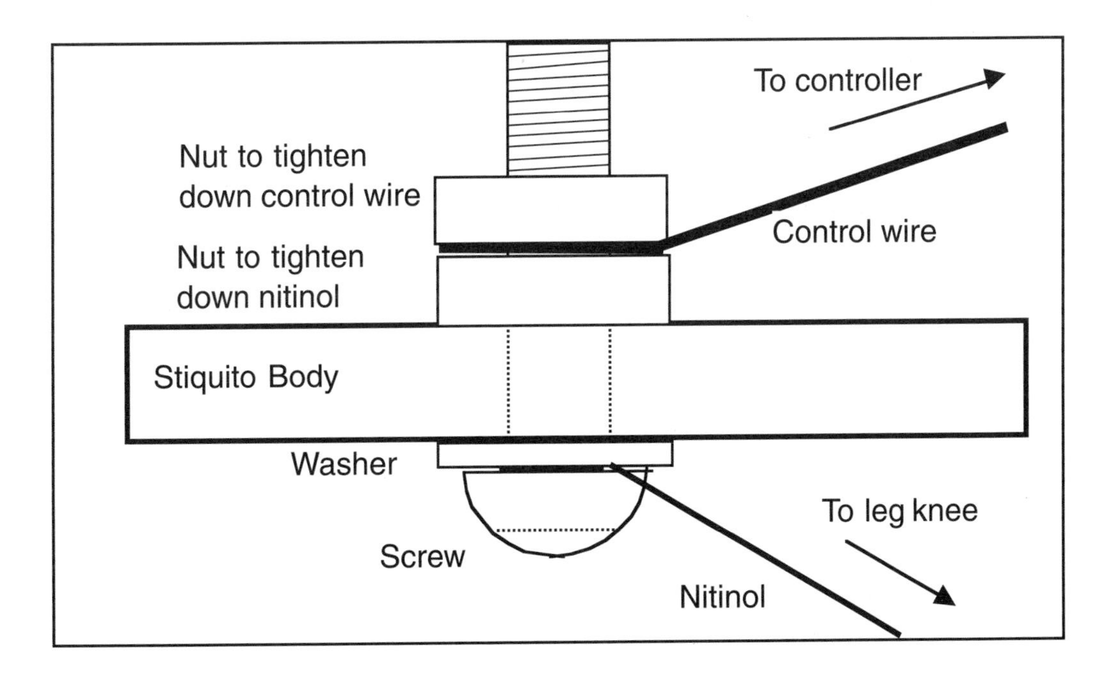

Figure AppC1. Using Brass Screws Instead of Aluminum Clips.

Index

P

Q

R

S

T

U

IEEE
COMPUTER
SOCIETY